P9-DEZ-538

The Ocean World of Jacques Cousteau

Quest for Food

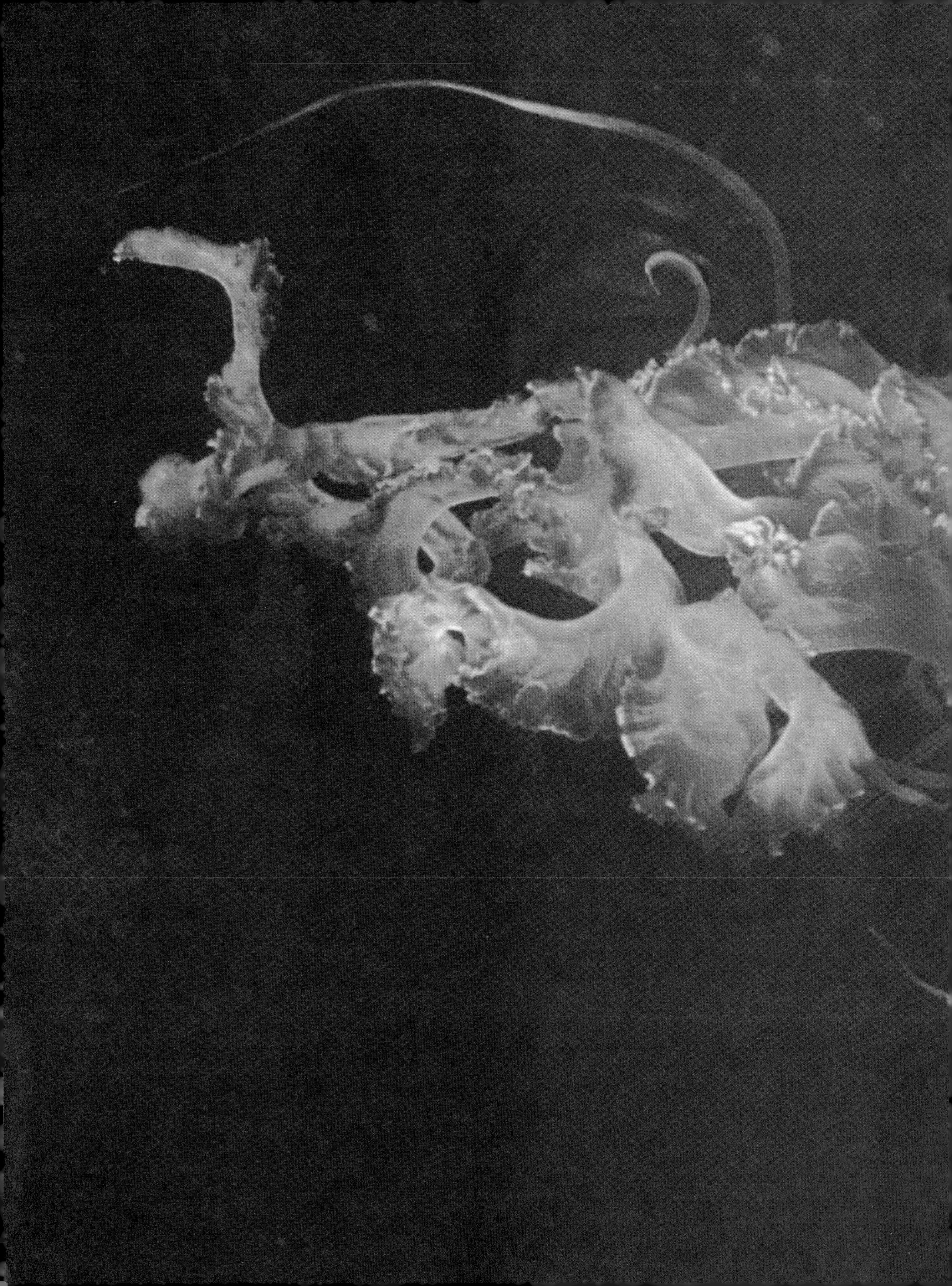

The Ocean World of Jacques Cousteau

Volume 3

Quest for Food

THE DANBURY PRESS

Sometimes evolution seems to be a race between a creature developing an offensive apparatus which can devour anything and a creature developing a defensive armament which nothing *can devour. Coral is as impracticable a diet as exists: hard rock sheltering tiny watery polyps and tiny algal plants. Yet the parrotfish has evolved jaws like beaks, with teeth turned into blades, capable of biting off chunks of coral and grinding them into powder. His digestive system then makes food of the algae and eliminates the pulverized rock.*

The Danbury Press
A Division of Grolier Enterprises Inc.

Publisher: Robert B. Clarke

Published by Harry N. Abrams, Inc.

Published exclusively in Canada by
Prentice-Hall of Canada, Ltd.

Revised edition—1975

Project Director: Peter V. Ritner

Managing Editor: Steven Schepp
Assistant Managing Editor: Ruth Dugan
Senior Editors: Donald Dreves
Richard Vahan
Assistant Editor: Sherry Knox

Creative Director and Designer: Milton Charles

Assistant to Creative Director: Gail Ash
Illustrations Editor: Howard Koslow

Production Manager: Bernard Kass

Science Consultant: Richard C. Murphy

Printed in the United States of America

234567899876

LIBRARY OF CONGRESS CATALOGING
IN PUBLICATION DATA

Cousteau, Jacques Yves.
Quest for food.

(His The ocean world of Jacques Cousteau; v. 3)
1. Marine fauna. 2. Animals, Food habits of. I. Title.
[QL122.C64 1975] 591.5'3 74-23064
ISBN 0-8109-0577-9

Contents

The quest for food is the strongest motivation for all marine creatures, and many millions of years of evolution have produced at random an unbelievable collection of weapons, strategies, and tricks. As is the case with all animals, TO EAT, TO LIVE is the rule of marine life. 8

For the most part the daily rhythm of the ocean features the alternation of FAMINE AND FEAST (Chapter I). Long stretches of quiet, unsuccessful search are followed, usually at dawn and dusk, by short but frantic feeding scenes. 10

FOOD IS NECESSARY FOR MAINTENANCE (Chapter II), but what use do the animals of the sea make of the food they take in. Each has a characteristic energy program. 18

To start with there are GROWTH NEEDS (Chapter III). Rapid growth immediately after birth is of substantial survival advantage in the sea. The baby seal is as big as its mother at three months of age. But most fish, unlike land animals, stop growing when deprived of food, and carry on growing until they die if they are able to find enough nourishment. 22

Next to the basic need to stay alive the most urgent drives are REPRODUCTIVE NEEDS (Chapter IV). Often an exceedingly difficult operation in the sea, reproduction can cost an animal a considerable slice of its energy budget. 32

Most sea animals have to move around—for a mate, for their own safety, for their dinners. Their PROPULSION NEEDS (Chapter V) require another big piece of their energy budget—in some cases the biggest. 40

The varieties of CONSUMPTION AND CONVERSION (Chapter VI) are almost countless. Most advanced are the carnivorous predators, who swallow highly concentrated food and process it through extremely efficient digestive systems. The sea cucumber, on the other hand, is a kind of mini–vacuum cleaner on the bottom—sucking up everything it comes across, sifting food from debris. 46

Marine animals spend a good part of each day LOCATING A MEAL (Chapter VII)—or letting the meal locate them! The lateral line organ helps most fishes; the dolphin has his built-in sonar apparatus; other animals possess eyes (some good, some mediocre), barbels, nostrils—or, in the case of the electric ray, a generator. 56

At the passive end of the scale we find SEDENTARY FEEDERS (Chapter VIII). Mussels and sponges simply attach themselves to a substrate, open shells or "mouths," and wait for food particles large or small to float in. A moray eel also sits and waits—but he only sits and waits until an unwary prey has swum just a bit too close to his lair. 68

The ACTIVE FEEDERS (Chapter IX) supply the drama in the life of the ocean. Rich in nutrients and varieties of plant and animal life, the reef communities are thronged by creatures actively in pursuit of a meal. An armory of weaponry enables the active feeders to hold onto and digest a prey. 84

Over millions of years FEEDING RELATIONSHIPS (Chapter X) have been established among some species. Animals as different as an anemone and a clownfish have learned to share between them the business of food getting in a symbiotic arrangement. Many smaller animals, like the Caribbean shrimp, perform cleaning services for much larger animals like the moray eel. Sharks are companioned by their pilotfish. 104

Man's relationship to the sea is largely that of a predator who takes what he wants from it and relies upon its natural fecundity for its replenishment. With PRIMITIVE FISHING METHODS (Chapter XI) there is nothing unhealthy about this. The primitive fisherman does not exterminate species or pollute huge oceanic regions. What he removes from the sea, the sea can easily, naturally replace. 120

But in FISHING THE SEA TODAY (Chapter XII) we find man playing the sinister role of parasite. Because of the explosion in earth's human population, and the misuse of man's progress, the demands he is making today of the ocean are exceeding its capacity to restore itself. 124

The answer lies in scientifically exploiting the resources of the sea, largely through MARICULTURE (Chapter XIII). Properly enriched and "farmed," an area of the sea's surface no larger than Switzerland could produce a larger annual yield than all working fisheries combined. 134

In the long run THE BALANCE OF NATURE cannot be overpowered or repealed. If a man is unwise in the use he makes of the world's resources, he may through scientific ingenuity put off the evil consequences for a year or a decade—but ultimately his folly will catch up with him. He will find that by destroying the productivity and fertility of the sea he has condemned his own civilization. 142

Introduction: To Eat, To Live

All kinds of prejudices encrust our thinking about the sea. Of no area, perhaps, is this truer than that quest for food which is so significant an activity in the animal world.

How many times, for example, have we read in books, in magazine articles, or heard in the conversations of swimmers and sailors that the shark is a killer? Quite a few highly successful films have been distributed which show closeups of large sharks in "feeding frenzies" —thudding into carcasses with robotlike persistence as hormones take over command of their mighty bodies, fearsome teeth and jaws scooping out great hunks of meat as easily as I scoop up a spoonful of soft butter, the roiling sea filthy with blood and carrion. It is certainly an appalling sight—quite enough to send chills down the spine of anyone, especially, perhaps, if he has witnessed such scenes firsthand.

A man's carnivorous meal also includes an ugly scene that is simply concealed and happens in the slaughterhouse or on farms where pigs or calves are bled to death. But a shark has no way to pretend innocence. A primitive animal in many respects, he possesses, to start with, only a cartilaginous skeleton, not a bony one—and there is nothing at all intimidating about his brains. But, as if in compensation for his rather lowly position in the vertebrate family, he has evolved into an epitome of muscle power and streamlined design. It is by exploiting these advantages, plus his capacious mouth with its row upon row of razor-sharp teeth, that he makes his living today as he has been making it for millions of years. He has absolutely no choice in the matter.

Still, it is not as easy as it might seem to rid ourselves of the kind of projection which turns this primitive, limited fish into a "killer," because our mental responses to the outside world fall under the influence of thousands of silent cultural monitors every second of the waking day. Most of us think of the sea otter as an adorable creature. Like the koala and the panda, its facial features are arranged in a pattern slightly reminiscent of a child's; there is something puckish in its expression, and our hearts spontaneously go out to it. On the other hand, even as notable and experienced a naturalist as William Beebe has testified to an involuntary spasm of revulsion upon seeing a large octopus slithering its way across a shallow tide pool. And the repellency with which the majority of mankind regards snakes may be something very ancient and deep in our genetic makeups—perhaps psychosexual in origin, or a dim memory harking all the way back to long-vanished ape-man ancestors.

The senses too become involved in all this. There are occasions, like the height of the season at a seal rookery, when the life of the ocean produces a stench all but intolerable to our nostrils. At other times, facing into a bracing breeze on a calm morning in temperate waters, a man blesses the day he was born and the fate which made him a seaman. There is no way of avoiding such subjective impressions. But we must control them as much as we can, lest —like the concoction of myths or false parallels between human manners and those of the sea—they disastrously interfere with our main business: to perceive and understand the dynamics of the ocean world. The sea is a huge and complex ecological machine in which every part interplays with every other, every process emerges from processes that precede it and feeds into processes that follow it, each death is the introduction to new life. To understand how it all works one must put aside everything in the way of sentimentality or personal bias. Then one becomes able to enjoy the beauty of any natural act.

What distinguishes living things from inanimate matter is the living organism's ability to appropriate materials from the environment and incorporate them into its tissues according to its own blueprint or, through metabolism, to break out from them the energies it needs for existence. In a sense, this is all there is to the topic of the quest for food. Different organisms live in different styles—and the life-style of an organism largely determines what sort of food it requires. A one- or two-celled animal in the plankton layers has next to no energy expenditure, so it can subsist on a thin diet of microscopic plants. A larger animal will have a larger energy budget; it may need its fuel intake in highly concentrated form and may be, as the shark is, a carnivore. The animals of the sea are never gluttons without reason. Never sadists, they eat to survive, as all animals must. They never kill for sport, they never torment their prey before killing it, they never exterminate whole species or render uninhabitable vast provinces of their living space. The shark is a most efficient carnivore, true. But he is no more a killer in the criminal sense of the word than the sedentary coral polyp which rapaciously ingests anything it can get its tentacles on, or the meathandler in a Chicago slaughterhouse—or, for that matter, the housewife who serves bacon at the family's breakfast table.

Jacques-Yves Cousteau

Chapter I. Famine and Feast

Some years ago, while *Calypso* was making her way through the Indian Ocean, we encountered a large number of sperm whales traveling in small groups with young. These are large animals; even the calves weigh close to a ton at birth. We had a terrible accident. A female swam across *Calypso*'s bow, and we could not avoid crashing into her. She would recover, but no sooner was she out of the way than a young whale in search of her, doubtless her child, ran into one of our propellers. The little whale was hopelessly cut up, bleeding profusely. To cut short its suffering, we killed it, then secured it to the boat.

"In warm seas, whales bleed profusely. The water about *Calypso* turned red. The spectacle began. Sharks appeared, more and more of them. When the frenzy began it was sudden and swift."

In warm seas, whales bleed profusely. The water about *Calypso* turned red. The spectacle began. Sharks appeared, more and more of them, attracted by the blood. For a time they nosed the baby whale, circling confidently, nudging and smelling. When the frenzy began it was sudden and swift. One shark lunged in and sliced away an enormous mouthful. The others joined. The animals would race along the length of the whale's body, taking chunk after chunk in quick succession like corn from a cob. There were more than thirty of these sharks. It appeared they had gone several days without a meal. But soon, as rapidly as it had begun, it stopped. The sharks had gorged themselves and could not finish all the food.

In less dramatic ways this pattern repeats itself everywhere in the sea. There is great deprivation, sudden abundance. In the lives of most marine creatures, feeding and the hunt for food occupy more time than any other activities. In some cases, when an animal has evolved a means for continuous ingestion, and its diet is always plentifully available, the feeding process is constant. At no point in its life does it stop eating and growing. The sea may hold an awesome abundance of food, but there is also an abundance of creatures to be fed at every level of the food chain. The competition for food is so fierce among most species that many animals must endure long periods of famine. When they are fortunate enough to find food, they feast.

Most fishes are cold-blooded and neutrally buoyant, so they need no muscular effort to fight gravity as land animals do. They are able to live well under these conditions. Their bodies adjust to the temperatures of surrounding water without the expenditure of energy. Their metabolic rates are so low that the energy they use in normal movement and respiration is minimal. Deprived of food, they lose weight just as humans do, but much more slowly—mobilizing first body fats and then proteins. This means that, even though they cannot grow without food, they can survive without food for a remarkable time in the open sea.

Some fishes, of course, are exceptions, with a high metabolic rate. The tuna is one. Because it has an enormous need for food, it is constantly on the move to find it, and because it is an exceptional swimmer, capable of bursts at 45 miles per hour, it quickly expends its energy, thus requiring more food to replace the loss and moving ever faster, rather like an ambitious and highly paid business executive. Though the tuna remains a cold-blooded animal, with all this activity its body temperature is always a little higher than the temperature of the surrounding waters.

Generally food is scarce in winter because of less growth of phytoplankton at the source of the food chain. The animal world

> "The competition for food is so fierce among most species that many animals must endure long periods of famine. When they find food, they feast."

depends on plants for its food, and the great abundance of microscopic plants in the sea provides the basis of a complex web of marine populations that support one another. Using the energy from the sun, plants absorb minerals to make basic foods. Grazing the plants, herbivores transform these simple materials into the proteins of their own bodies, comprising the food source for carnivores higher in the chain.

In the summer months, when more solar energy is available, a greater volume of phytoplankton grows, providing a greater volume of food. This not only causes the individuals of a species to grow to greater size. But it also increases the supply of food at each level of the food pyramid, allowing an overall increase in the mass of living organisms. Populations of the large predators high in the food web may rise as a result of more phytoplankton supporting a greater supply of food.

When we begin to appreciate the apocalyptic scale of consumption that takes place in the sea, vast populations feeding one another, each creature a threat to another creature and/or threatened in its turn, we may picture the water environment as a kind of hellish bedlam of torn flesh and starving predators. It is not so. Except for the hours at dawn and dusk, when the sea boils with feeding activity, it is for the greater portion of each day as calm and composed as you see it here. Feeding is going on, but it is the quiet filter feeding of such animals as oysters, barnacles, feather-duster worms, and coral polyps, which, fixed in place, wait tirelessly for food to drift by.

Passive feeder. *The sea fan, pictured below, is a colony of animals attached to a rock, waving its wide lacework fan according to the sea's surge, awaiting its meals. Eventually plankton drifts by. Then the sea fan can eat.*

The Changing of the Guard

Twilight is the time of day when the sea bubbles with a frenzy of activity and excitement. Like day-shift workers, diurnal animals prepare to retire while nocturnal animals rouse themselves and look for food. Sunlight plays an important role. Sunrise and sunset are feeding times for most of the animals in and around the sea. The nature of the light at these times of day makes fair game of fishes usually protected by countershading. Now with sunset, the slanting rays of light silhouette all these creatures and predators below see dark forms everywhere above in clear outline. For a few, often fatal moments, the protective coloration fails.

> "Sunrise and sunset are feeding times for most animals in and around the sea."

Gulls at sunset. *As the day ends, gulls flock close to the water to catch their evening meal. With the change in light, they can more easily see their prey.*

Fish do not usually remain in schools at night because schooling depends primarily on vision which is reduced at night—and to do so would make them that much more visible to attackers. Most schooling fishes separate, individuals abandoning the mass and swimming away in all directions to find protection in the dark. They are most vulnerable in these moments of transition between day and night. Yet the first nocturnal fishes are also prey to diurnal predators whose peak feeding occurs at that time. Soon the light-shy creatures of the deep scattering layer will begin to rise to feed on the plankton, with them the animals that prey on them—and those that prey on *them.* We have never been able to photograph feeding time underwater. As soon as a diver slips into the water, all activity stops.

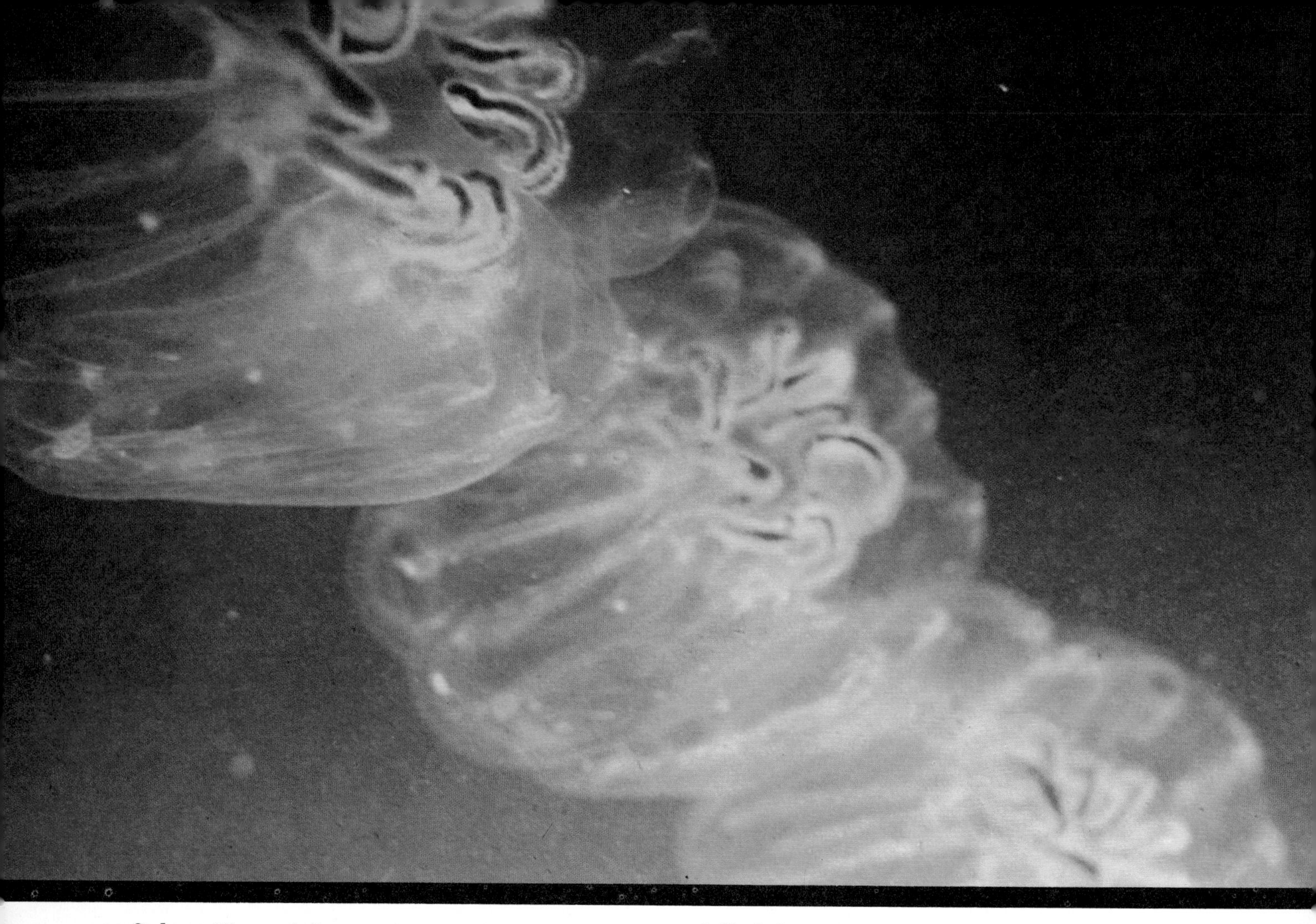

Salps. *These delicate, almost substanceless animals move gracefully in the sea, finding their food which is made up of plankton. Each propulsive contraction entails the ingestion of food.*

Jellyfish. *These animals, depending on the species, capture and eat plankton, eggs, worms, larvae, and small fish. They capture them by means of their stinging cells, and hold them with a sticky mucus.*

Constant Feeders

In the open sea food is scarce. Predators must constantly search for a meal. In some circumstances an animal is equipped for constant feeding, and its diet is continuously available. The salps enjoy both of these conditions at once. Contracting and expanding to push water through their hollow middles, the salps move forward by the jet of the water expelled behind. As the water passes through, they filter food along their mucous membranes and excrete wastes. A single rhythmic motion, which continues as long as the animal lives, accomplishes both feeding and propulsion. The jellyfish may not always have food at hand, but they are always ready to respond to its touch. Stinging cells react instantaneously to paralyze the food, which is then taken up to the center of the large float.

Jellyfish vary greatly in size. Some of the largest have bells eight feet across and tentacles extending downward to 200 feet. The bulk of their bodies is made up of a nonliving secreted jelly which fills two layers of body tissue and gives the animal buoyancy. In the more advanced ones the jelly has been strengthened by cells and fibers, but even the strongest are more than 94 percent water. Muscle tissue of a jellyfish totals only about one percent of its weight.

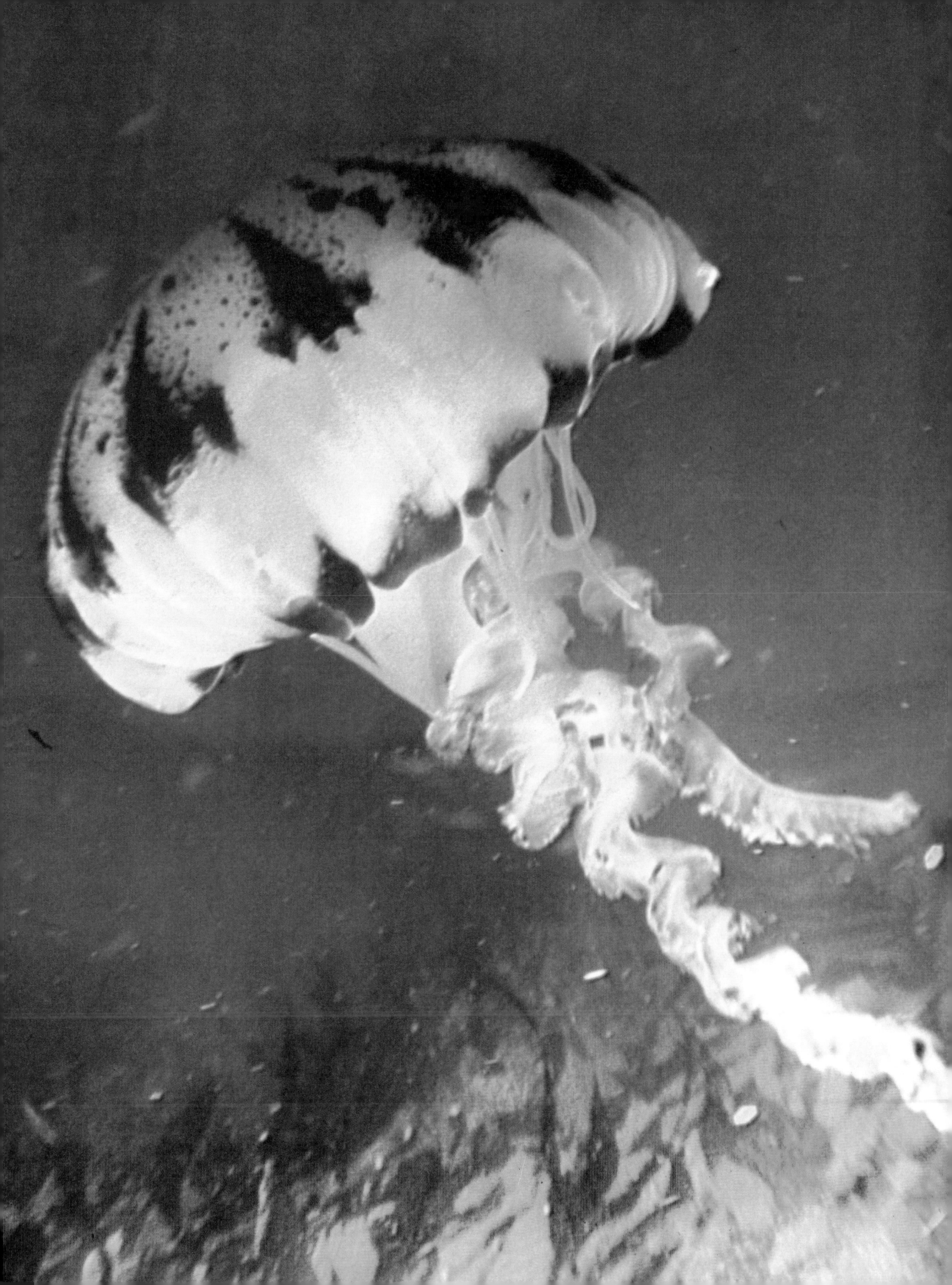

Basking shark. *Just smaller than whale sharks, the largest of all fish, these creatures consume huge amounts of plankton. When the plankton supply is down, the basking shark hibernates.*

Food Scarcities

As sunlight wanes in winter, the growth of phytoplankton slows. The resultant scarcity of this basic food source is felt throughout the marine chain. Many members of the sea's population move on to more fertile areas when the seasons change, others drastically alter their feeding habits. Some herbivores may become carnivores. Some animals stop eating entirely for many months.

The basking shark is a plankton feeder. Its gill apparatus is lined with a thick mucous membrane to collect the plankton borne by water as it filters through. In the North Sea the density of plankton has dropped by November of each year to a point where this animal, merely to collect its food, must expend more calories in an hour than it can replace in the same period of time. So the basking shark becomes a dropout. It gives up work entirely, sinks to the bottom, hibernates—shedding its old set of gill rakers and growing new ones for the spring, when it will rise to the surface and begin to feed again. Cratered on the ocean floor at vast depths, all its muscles relaxed, its body systems dormant, its heart only faintly beating, this giant lives without food for months. Our automatic deep cameras have often photographed mysterious excavations in the mud

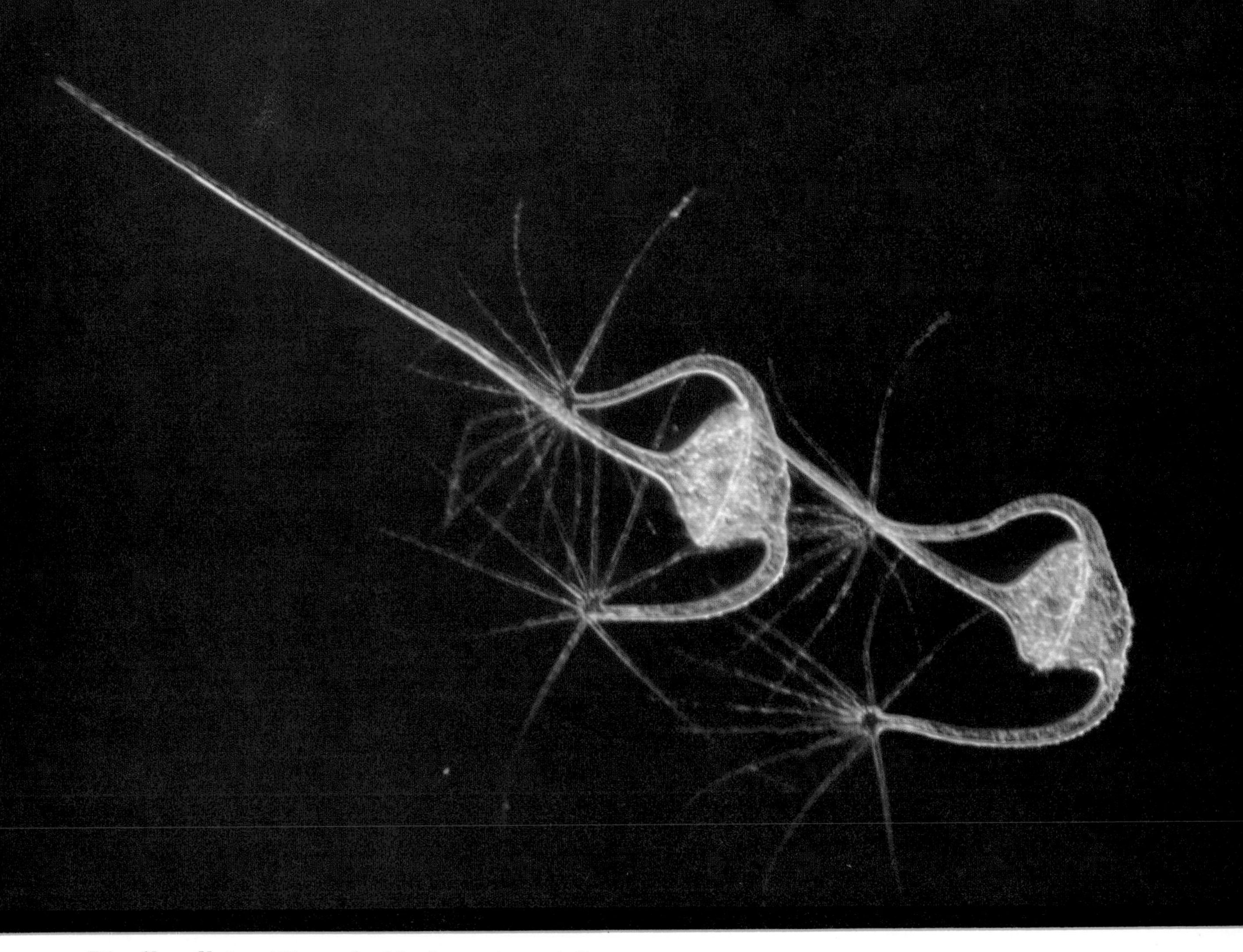

Dinoflagellates. *These planktonic creatures gather nourishment from whatever source they may. They can use the sun for photosynthesis like a plant, or make use of other material like an animal.*

of the ocean floor, at depths as great as 8000 feet. Large numbers of ruts and pits are found close together. It appears that huge objects have settled there. We now believe these are the marks left by basking sharks, who may settle together in great sleeping colonies to pass the winter months.

Still more remarkable is the versatility of some dinoflagellates in the plankton. For energy and growth, the same unicellular creature may carry on photosynthesis, like a plant, or utilize organic material, as an animal does, according to whichever is best suited to the food at hand. Other plants and animals are able to take maximum advantage of the limited food supply by alternating life-styles, so exposing themselves to food in more forms than would otherwise be accessible to them.

Animals like the salmon and the common eel (*Anguilla*) impose the hardship of famine on themselves, going without food for months during migration to their spawning grounds, living off the fat cells stored in their muscle tissue until these are totally exhausted. The animal's system can rarely recover after such treatment, but so strong is its instinct for reproduction that nothing interrupts its martyrdom. Less drastic in their fasting during reproduction are the seals and whales that go to gentler climes to mate and give birth.

Chapter II. Food Is Necessary for Maintenance

An animal is a complex creature. Aside from its capacity to grow, it feeds, digests, respires, excretes, forms eggs or sperm, repairs its damaged tissue, and in most cases orchestrates a set of muscles in order to move in various ways at various speeds. All these functions most animals perform every day. They must do so or perish.

To keep the intricate array of its body machinery in constant running order, an animal uses a variable amount of the energy it derives from food. A fish, being a cold-blooded animal, adjusts its body temperature to that of the ambient water, and thus has practically no heat loss—and heat is the most costly form of energy. Another advantage a fish enjoys over land animals is that, when he is not swimming, his body is fully supported by the water's thrust, so that he does not have to use his muscles to fight gravity as we do all day long. For all these reasons, most marine animals (except mammals) can maintain their lives with extremely small needs of energy—and consequently very small food requirements. Most of their energy budget is devoted to growth, to propulsion, to sex, and to regeneration of damaged parts of their bodies.

The efficiency of the process of assimilating energy from food varies from one animal to the next. Not everything an animal eats can be converted to energy, and some foods offer a higher yield of energy by volume than others.

Some animals have more extravagant needs than others. The more simple its system the less a creature needs to keep itself alive. Seeds and spores, many of them unicellular structures, can remain dormant for years and still spring to life at the end of that time. Their energy demands are minimal.

The most expensive demand an animal can make upon its energy supply is a constant body temperature. This is achieved by warming the blood to a fixed level, which provides the animal with an internal environment at optimum temperature for the highest efficiency of its thermodynamic machine. All mammals have this costly feature. It is a great asset. The animal is capable of sustained efforts and endurance several times greater than cold-blooded creatures of the same size. Marlins and sharks can beat a dolphin in a 100-yard race, but cannot compete at cruising speed. Tunas and some sharks have internal temperatures several degrees higher than the sea they inhabit. They pay dearly for this, burning energy at such a high rate that they have had to become constant feeders.

> "The energy budget of most marine animals is devoted to growth, to propulsion, to sex."

As an animal grows larger, its surface area increases, but not so quickly as its total body mass. A large mammal, therefore, has a smaller surface area in proportion to its mass, and thus has relatively much smaller losses of heat. This makes the task of maintaining a warm body easier than for smaller creatures. It accounts for the great size of the polar bear. Less energy is spent in heating each unit of volume in the polar bear's body. Less fortunate are small mammals like the sea otter, whose body is small, bringing most of its mass in close proximity with its exposed surface area. The sea otter works very hard at keeping its body warm and can only achieve this by constantly rubbing air into its fur for insulation.

Regeneration

You are observing a rather special process which can tax the energies of members of the starfish family. This one is regenerating an arm it somehow lost. The ability to regenerate parts is the legacy of a relatively simple nervous system. As evolution moved into the vertebrates, with their complex nervous and somatic systems and articulated skeletons, the power to sprout new limbs in place of lost ones was one of the prices paid. But in some invertebrates it can be carried to astonishing lengths. There is a species of starfish in Australia that can regenerate itself completely from a small piece of a single arm. In other varieties at least part of the central body disc must be involved. The starfish also seems to be one of the longer-lived animals. Some live to 20 years.

Warm and Cold Blood

When an animal enters the water, the heat drain becomes severe. In some animals that live in the sea, like the polar bear who wallows comfortably in icy arctic waters, a combination of their fur and fat serves as protection. We saw that the sea otters had little fat and fur on their spare bodies when we were filming them off California for television. We saw how they kept eating during most of their waking hours to maintain not only their body temperature but also their weight. To do this, we learned, they must consume up to 20 pounds of abalone, crabs, and sea urchins a day. That's equal to one quarter of their entire body weight. By contrast, a man living on land eats little more than 1/100th of his body weight each day.

Cold-blooded animals are another story. Fishes and reptiles that live in the sea don't have a constant body temperature to main-

tain. Their body temperature varies with the water they swim and live in. Sometimes it's a bit higher than the surrounding water. Occasionally, as in the swift-swimming tuna, it's substantially higher. Then a tremendous amount of food-supplied energy is needed. But the tuna is an exception among the cold-blooded animals. Most are capable of going without food for days, weeks, or even longer periods. Warm-blooded animals, on the other hand, must fuel their bodies regularly and in quantity or perish. One of the greatest perils facing the aqualung diver is fatigue brought on by the drain of heat from the body.

"Sea otters must consume up to 20 pounds of abalones, crabs, and sea urchins a day. That's equal to one-quarter of their entire body weight."

Underwater visit. *A diver, visiting the remains of an old shipwreck, discovers that it has been colonized by various forms of marine life.*

Chapter III. Growth Needs

Everyone's heard of the runt of a litter, the pitiful pup or kitten smaller than his litter mates. What causes one animal to be feebler than his siblings?

One of the principal reasons is poor nutrition. If a pregnant mother is poorly nourished her children will be too. In multiple births, common among lower animals, if one of the embryos is crowded inside the womb, or if its connection to the mother is twisted or partially blocked, the result is a runt. Anytime a young animal in its critical growth period is starved or otherwise improperly nourished, the result is stunted growth. If you have seen pictures of children in Biafra or Bangladesh, you know this. The scrawny child of ten looking like four is a victim of poor nutrition.

These negative examples provide us graphic proof of the importance of good food in quantity and at a time when the body needs it most. While this is true in mammals, where a complex hormonal mechanism stops the growing process when an individual reaches the adult state, the same does not necessarily hold with fishes. Here growth is a continuing process throughout life. Nutrition plays a major part. Without proper amounts of the appropriate foods, fishes will temporarily stop growing. But, unlike mammals who suffer permanent and irreversible damage from malnutrition, fish may resume growth and overcome the effects of insufficient food. In mammals, growth may resume when proper nutrition is resumed but there's no way of making up for lost growing time.

> "Anytime a young animal in its critical period is starved or otherwise improperly nourished, the result is stunted growth. If you have seen pictures of children in Biafra or Bangladesh, you know this."

Among the lower animals, those with no backbone, many are affected only slightly by starvation periods. Mosquitoes can pass through two or three generations without eating—at least in their adult stages. Yet their young, the larval wrigglers, are unaffected. Others, like sea anemones, shrink and grow depending on food intake.

One of the most unusual groups of animals is the sponges. They feed by waving hair-like cilia surrounding their feeding pores. The cilia wave nutrient-laden water in, and special feeding cells take the food and utilize it for the good of all the various cells in the sponge. But the sponge can be strained through a piece of cheesecloth so that each cell is completely separated from every other cell of the sponge. When all the cells are put in a container together, they'll rejoin and resume growing.

Oxygen is as important as food. As the body takes it in, either via the lungs or via the gills, oxygen travels through the circulatory system to all the cells of the body. Oxygen is an important part of the chemical process which breaks down food and yields the energy necessary to all body functions.

In the process of growth, food is broken down chemically into basic units of carbohydrates, fats, and amino acids. These will be further reduced to simple compounds for energy production, or they may be incorporated into the cells and tissues directly. The latter process is essential to provide substances that the animal cannot synthesize.

Rapid Growth

This baby seal must take in enough food to grow up in a hurry. The colony meets for about two months on a peaceful arctic island, or a gigantic ice floe as large as an island, to give birth to one year's offspring and procreate the next. Then it is the open sea—with all its rigors and dangers—for every member. Some seal pups are able to swim well enough almost immediately, but others stay on land or ice for two or three weeks before they are introduced to the water.

Usually the mother seal returns to the sea a week after the pup is born, but the little one is not thrown completely on his own. His mother continues to come back to suckle him once a week for three to six weeks. But after that she cannot catch his fish for him or protect him from the dreaded orca, whose favorite tidbit he is. The pup must grow to nearly the size of his mother in about three months to be able to face the challenges alone.

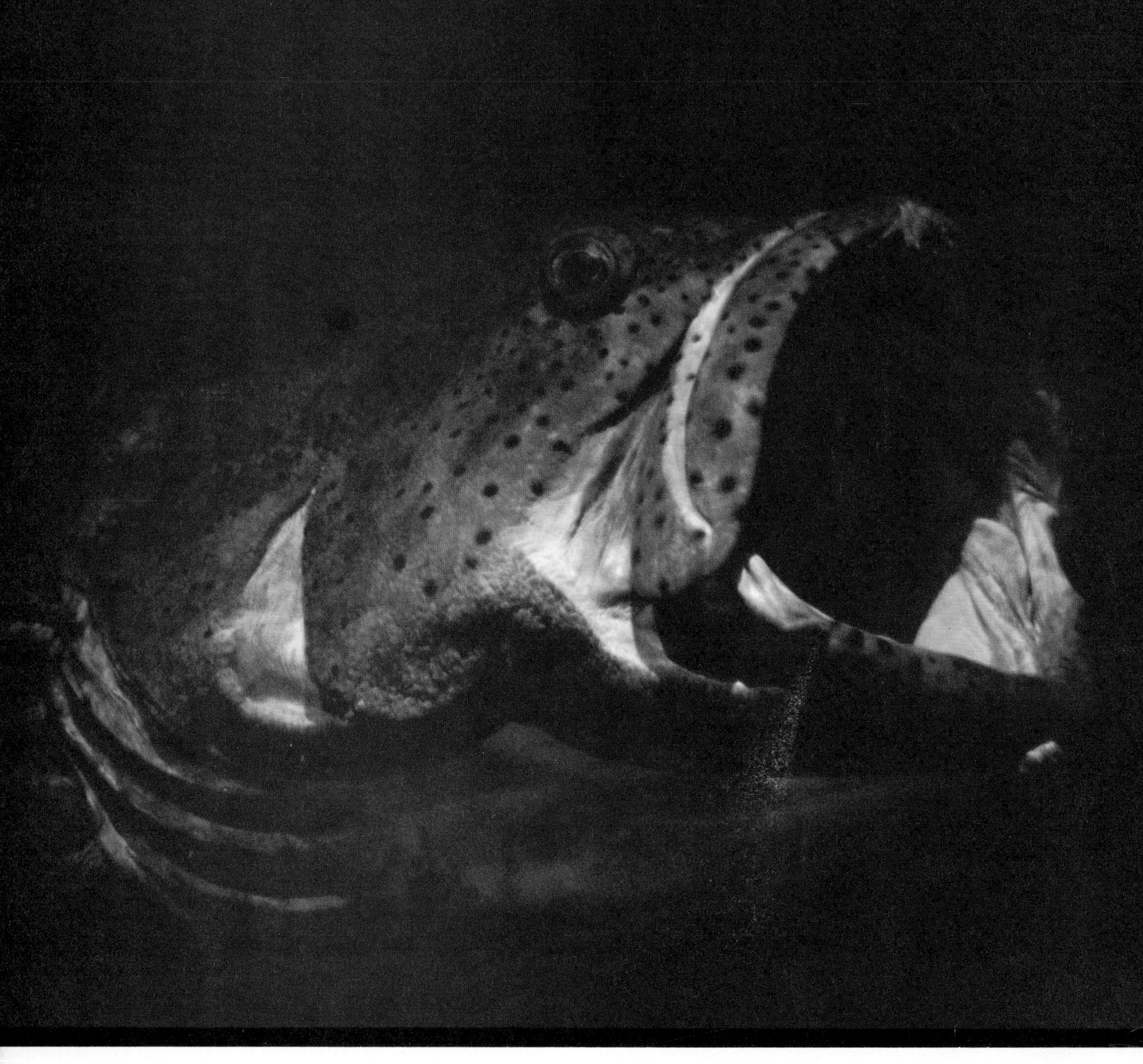

Jonah—and the Grouper

It is questionable whether or not the biblical story of Jonah and the Whale has a kernel of truth in it. But if long ago a traveler was thrown off a ship and "swallowed," it could not have been a whale that did it. But there are other large animals in the sea. Once, investigating wrecks in the Red Sea with *Calypso,* our group encountered an enormous grouperlike fish some 13 feet long. We had never before seen such a gigantic fish of this kind. Groupers inhale their meals. So, what could have happened in the biblical past is this: Jonah fell or was thrown overboard, he was inhaled by a grouper, temporarily lost consciousness, and was immediately exhaled by the fish, luckily close to land. Here is an experience no protagonist would cease talking about for the remainder of his life—and perhaps it is this wild ride which the world has never forgotten.

Growth Rings

Many fish in temperate waters grow more rapidly in summer when food is plentiful than in winter when it is scarce. The record of growth is printed on the hard parts of fish—scales and such bones as otoliths in the inner ear. Fish that have scales only begin to get them when the fish are about an inch long. They appear first at the base of the tail and spread in overlapping rows to the head as the animal develops. The number of scales on an individual fish remains constant during its lifetime, for they grow as the fish does. Each new size change is marked on the scale by ridges or growth rings, bunched together in winter months, spread out in summer when growth is rapid. In those species that change from fresh water to the open sea, the sudden spurt in growth that comes with their new habitat clearly tells us how old they were when they made their move.

Lobster's Arsenal

Lobsters may feed on living animals, but they are basically scavengers. Because they need calcium for their external skeleton, they will consume sea shells and may even engage in cannibalism. The rigid shell cannot accommodate the growing lobster inside. So every few months, after the lobster has grown about 15 percent, its armor coat must be shed. Then a new shell must be fabricated, which requires more food. Growth is no simple process for those with their skeletons on the outside. The two big claws each serve a separate function in feeding. One is heavy with large blunt teeth on the inner edge. It is the crushing claw. The smaller one has fine, sharp teeth. It is the tearing claw.

Varying Growth Rates

Cold-water animals are usually larger than warm-water ones of the same species. One explanation given by scientists is that in cold water animals need more time to reach sexual maturity and therefore have more time in which to grow. Oysters grown in the Cape Cod area of the Atlantic coast, for example, do not spawn until they are two years old. Those in the Gulf of Mexico may spawn at the end of their first summer—when they're three to four months old. In some cases populations of island fishes do not attain the length of the same species found along the mainland. This is true for a tube blenny, which is smaller around Cocas Island than along the Mexican mainland.

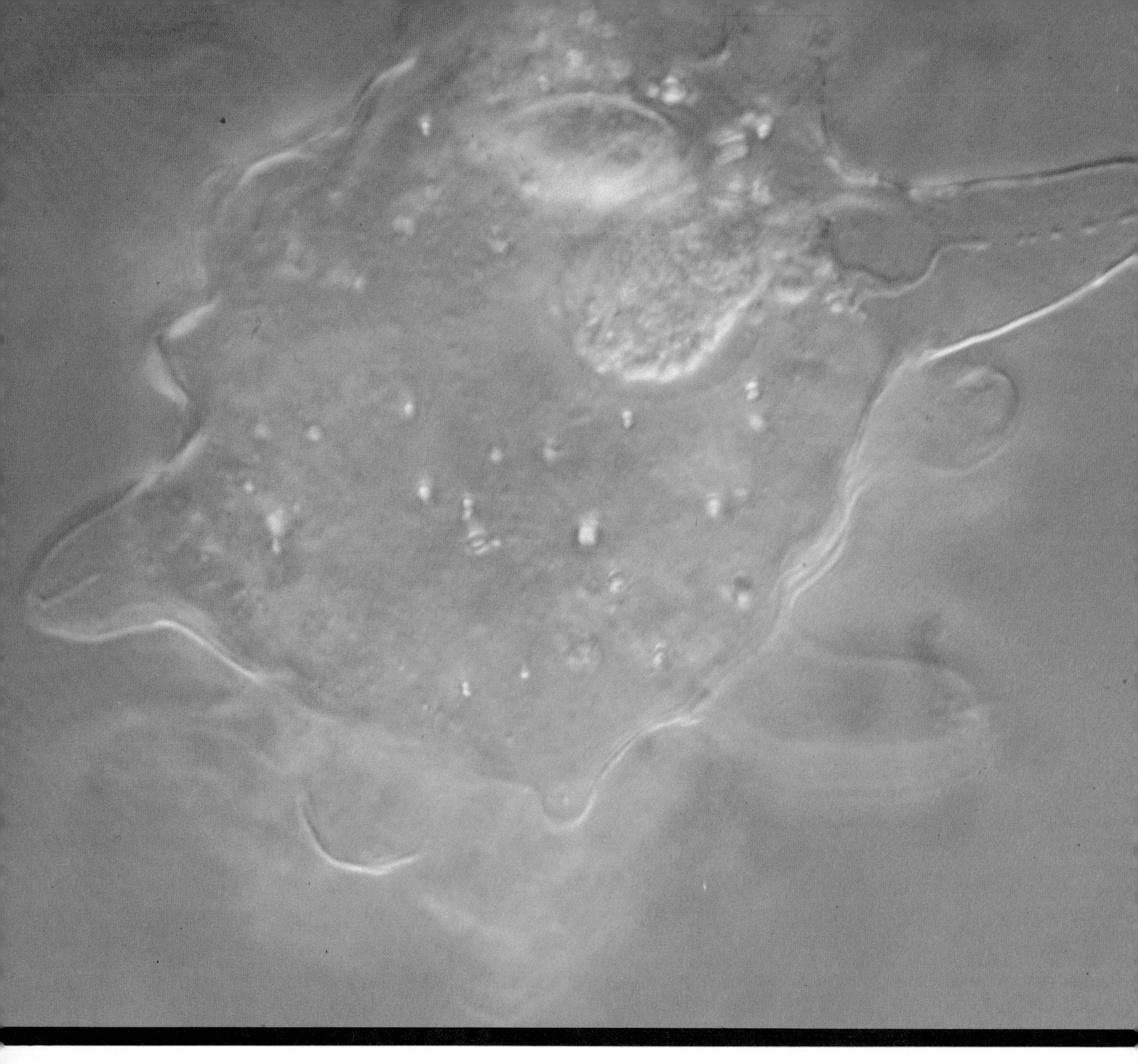

The Discerning Amoeba

All forms of life in the ocean, from the microscopic amoeba to the giant whale, need food to survive. Their diets are inseparably tied to their food-getting ability. The one-celled amoeba feeds on plants and animals smaller than itself: bacteria, yeasts, algae, other single-celled animals. Although one of the simplest forms of life, the amoeba has the power of distinguishing between grains of sand or other indigestibles—and food fit to eat. It does this by sensing the change of chemistry the food causes in the surrounding water. The amoeba eats in different ways. Should the prey be a tiny mote of green algae, the amorphous animal extends itself around, envelops it completely, and ingests it. If it seeks to take a smaller protozoan than itself, it will send out long flowing pseudopods. Once the prey is trapped, the amoeba draws it into its food vacuole.

Disappearing Whales

The largest creature ever to live is the blue whale—now nearly wiped from the face of our planet. Whales of many species range the oceans of the world from polar seas to tropics. In the Canadian arctic the white beluga whales shown here are camouflaged when they surface in ice-filled waters with snowy backgrounds. But Eskimos still hunt and kill them, mostly for food. Another whale hunted by hand harpoon and rifle by Eskimos is the narwhal with its single long, helical "corne." Other species of whales—many of them great, harmless, toothless creatures feeding on plankton—have been systematically decimated by man for food, fuel, hide, bone, for art's sake and for corset stays. Many species of the small beaked whales and even some species of the closely related dolphins are so rare few men ever see one in a lifetime at sea.

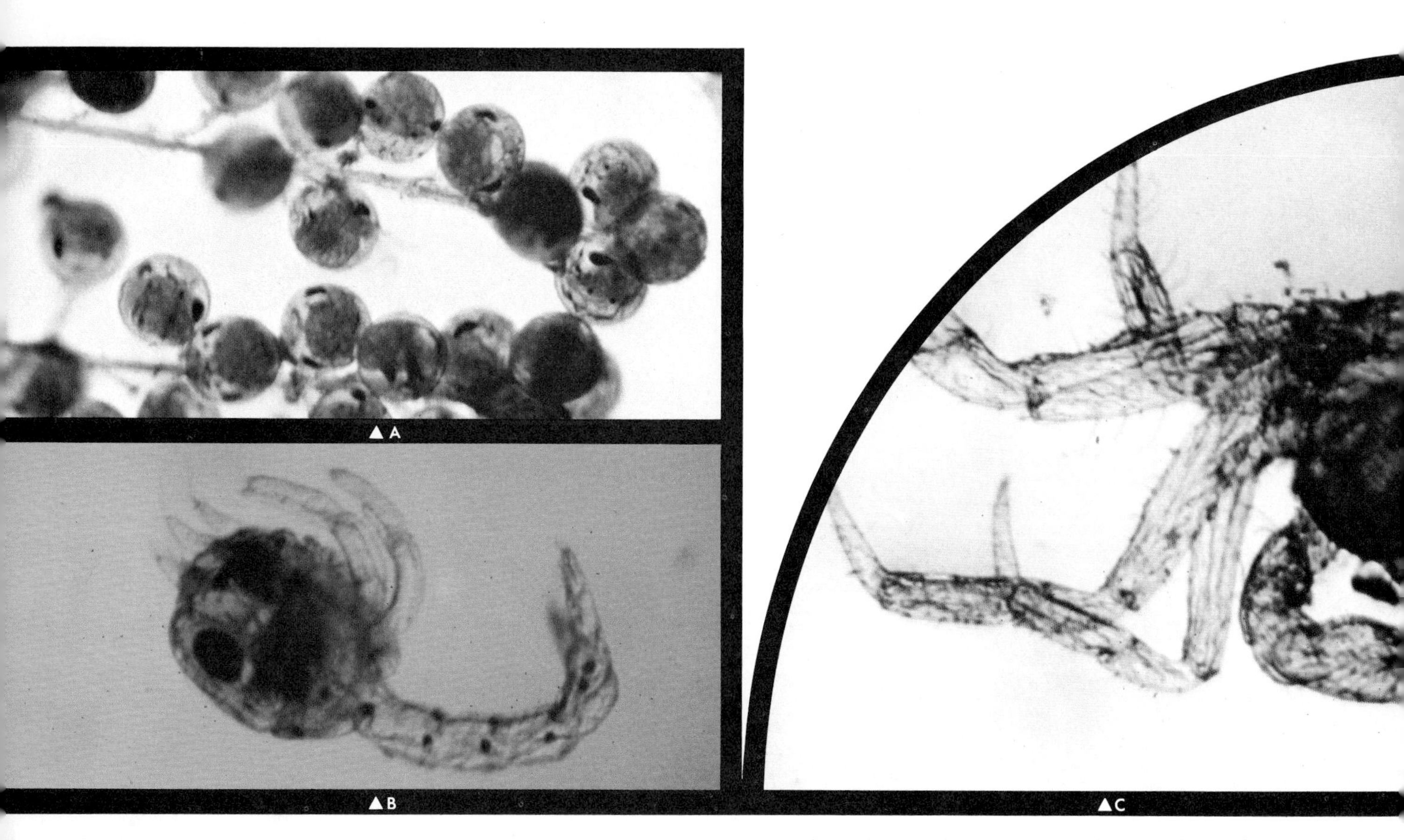

Ontogeny of a Crab

All animals pass through stages of development from egg to adult. Some go through few. Others, such as crustaceans, change many times.

Among the true crabs five distinct steps lead to adulthood. Inside the roundish egg the newly developing crab passes through its first phase, nauplius, during which it is egg-shaped and unsegmented. The broad front end carries on it three pairs of appendages—the antennae, the antennules, and the mandibles or jaws. It has a single eye in the middle of the front end but otherwise only a slight suggestion of the other organs and body parts it will have as an adult. The next stage, at time of hatching, is the protozoea stage, followed by the zoea stage. In both of these phases the crab begins to take on more of its future shape—development in some species of the hind legs, the beginning of segmentation of the abdomen, the start of a chest shield. The step before adulthood, megalops, brings the crab to something close to its eventual body shape.

In the early stages larval crabs are at the mercy of ocean currents—drifting about aimlessly, subject to the attack of larger plankton and fishes. In their turn, larval and adult crabs get hold of smaller plankton. To pass through these stages of development, changing form drastically in some of them, a vast amount of energy is needed.

> "Even in the metamorphosis of a human from fertilized egg to birth there are tremendous changes of form from tiny tail-bearing fetus with gill slits to newborn infant."

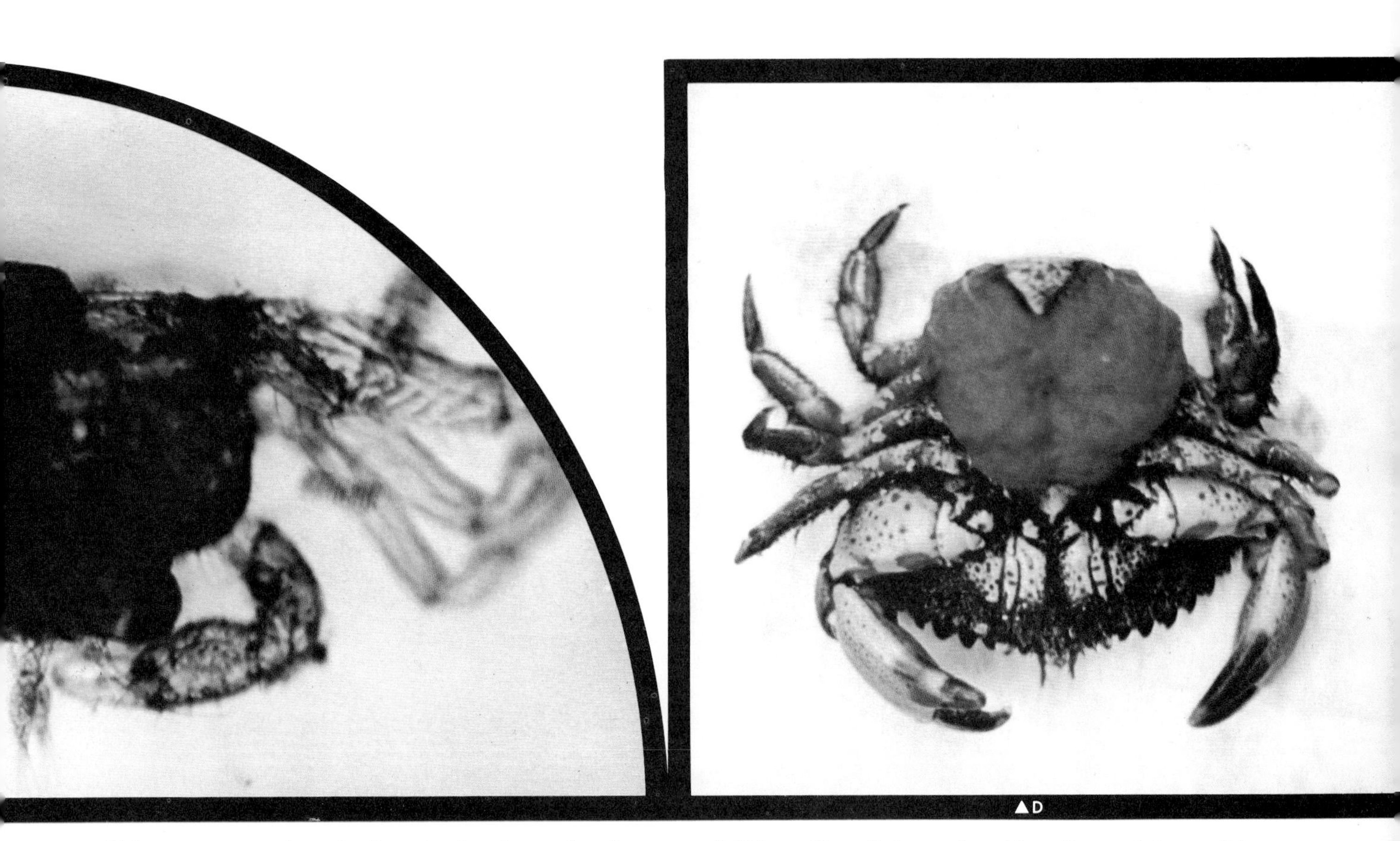

Thus, we can often judge the food needs of an animal by knowing how severe the changes he experiences in growing to adulthood. Among fishes, flounders start life as eggs. When they hatch they swim with their bellies down and their back toward the water surface above. And they have an eye on either side of their heads. When they get to be about one millimeter in length they flop over onto one side and live the rest of their lives on that side. When they first begin their lives on one side they have only one functional eye. So the eye on the side facing the bottom migrates over the top of the head to the upper side. This eye migration gives the bottom-dwelling flounder far better vision.

Even in the metamorphosis of a human from fertilized egg to birth there are tremendous changes of form from the tiny tail-bearing fetus with gill slits to the newborn infant. Some other animals pass through free-swimming stages before settling down to a sedentary life. The barnacle swims about freely in its nauplius stage, passes through a cypris stage, and finally fastens itself to a rock, metal, or wood substrate with what may be the world's most perfect glue. There it remains for its entire adult life, kicking particles of food into its mouth with a fanlike foot.

A / Nauplius. *Prior to hatching, the crab bears little resemblance to the animal it is soon to become.*

B / Zoea. *In this phase we can tell that the animal is going to be a crab—we can see its legs and the beginnings of its "breast plate."*

C / Megalops. *Except for the fact that many portions of the animal's body are still soft at this stage, it is unmistakably a crab.*

D / Adult crab. *Looking back over it development, we can now see where the crab's five pairs of limbs and carapace come from.*

Chapter IV. Reproductive Needs

We have seen in *The Act of Life* that in the perpetuation of most species in the sea there is a great loss, necessitating the production of thousands of eggs so that one or two will survive. In order to be able to build within their bodies, out of their own substance, so many life germs, the parents need to store energy for long periods of time, which means that the mating period is often the climax of a year or more of healthy, active feeding behavior.

Watch a male seahorse extruding miniatures of himself from his brood pouch and you'll sense the effort he must make. For minutes on end, in a series of spasms, his body ejects the baby seahorses he has cared for after receiving them in virtually embryonic form from his mate. For among seahorses, the female conceives and forms the young ones, then turns them over to the male for development. The tiny seahorses transfer into his brood pouch on the front side of the beginning of his tail. There they take nourishment from the male and grow stronger, so as to have a better chance among the grasses of the shallow seas.

The salmon, of course, pays the supreme sacrifice in reproducing its kind, dying days after its harrowing struggle upstream to spawn. In the five years or so which the salmon spends in the open ocean, he grows into a big, powerful fish with substantial energy resources packed into his frame in the form of muscle and fat. When that last spring comes, and he feels the reproductive summons, he needs every calorie in this storehouse. The navigational feat he performs in reaching the right freshwater tributary from the trackless wastes of the deep sea is itself fabulous; we still do not know exactly how he manages it. As soon as he leaves salt water he ceases to eat. Now, living solely upon his reserves, he commences the brutal trek upstream to the precise pond where he began life—there to spawn in his turn. The salmon accomplishes his last journey with such a prodigal expenditure of energy that at the end, after he has completed the reproductive chore, he is a decrepit caricature of his marine self. Terribly injured, his body tissues depleted, he floats off and dies.

"For minutes on end, in a series of spasms, the male seahorse's body ejects the baby seahorses he has cared for after receiving them in virtually embryonic form from his mate."

Even among those fishes that don't have to negotiate raging cataracts, or bump ungracefully over sandy beaches in shallow surf, a vast amount of energy is needed to create and supply new life. The sperm the male spills, the eggs the female drops are products of their own bodies, fed by the fuels of plants and animals. In the live-bearing fishes—sharks and cichlids and guppies, for example—and in the marine mammals like whales, dolphins, seals, and sea lions, the female must produce from her own body and give the growing embryo within her that which might have strengthened her. At the same time she may have to produce more to maintain her body so she can continue to feed and care for her young after they are born.

Salmon challenging a waterfall. *In the course of fighting his way back to his spawning grounds, the salmon expends just about all his energy. When his instincts tell him his time has come, he begins the difficult journey home. Once there, his mission accomplished, he simply gives up—and dies.*

Little Rising and Great Rising

The bristle worm, above, inhabits coral reefs and is a member of the annelid phylum, which includes the lowly earthworm. One member of this group is famous for spectacular body changes associated with their reproductive processes. One of the most curious is the palolo, a fairly big worm living among the coral reefs of the South Pacific. It reproduces every year in the last days of October and November—the "little rising" and the "great rising." At this time the worms are sexually mature; each individual has concentrated either eggs or sperm in the rear half of its body. This half develops its own eye; it separates from the front half, swims to the surface, and mates with a halfworm of the opposite sex. The parent-half regenerates itself for next year.

Storing Up for the Winter

During December and January of each year the California gray whale swims south along the coast of western North America from the Arctic Ocean to Baja California. It is moving into warmer waters to calve. What is interesting is that during the course of this journey pregnant female whales eat practically no food. They don't need it. Through long summer days they have gorged themselves on the krill-rich waters of the far north. They now have coats of thick layers of blubber, containing more than enough energy for the long reproductive migration. Once arrived in the subtropical waters of western Mexico, the mothers give birth. The pups nurse on the female's milk, building up the blubber, gaining the strength they need when the northward migration begins in March.

A Tough Bird

The penguin doesn't fly, and lives mostly in frigid locales. It's not the easiest life for a bird. The penguin is a swimmer and diver and water-skipper—as fast and adroit, it is thought, as seals and porpoises in pursuing the fish, cuttlefish, crustaceans on which he feeds. Since they are birds, penguins must find places suitable for hatching eggs when the time comes for reproduction. They make crude nests lined with what grasses or reeds or rocks can be collected to protect them. Two species of penguin, the king and emperor, give more elaborately of their energies to the next generation: they possess flaps of loose skin on their legs, which form a sort of pouch into which the single egg the female lays is slipped during the incubation period. In the case of the emperor penguin this all goes on in the midst of the coldest portion of the year.

Life Must Go On

The most important function of any animal is the perpetuation of its species. The amount of energy needed for this will be spent, whatever it is. The salmon whose eggs are being "milked" for incubation in the picture above spawns once in its life and within a week will die, spent by the life-long energy drains of preparing to spawn and of spawning. Other fishes can spawn many times in a single season and bring forth eggs year after year for decades. In a ripe female salmon, the eggs may be as much as 10 percent of the fish's total body weight. The gonads in a male salmon weigh 5 to 7 percent of his weight. In a 450-pound bluefin tuna the eggs of a ripe female equal about 5 percent of her weight, while in the striped bass, with its larger and more numerous eggs, they account for close to 15 percent, or two and a half pounds in a 20-pounder.

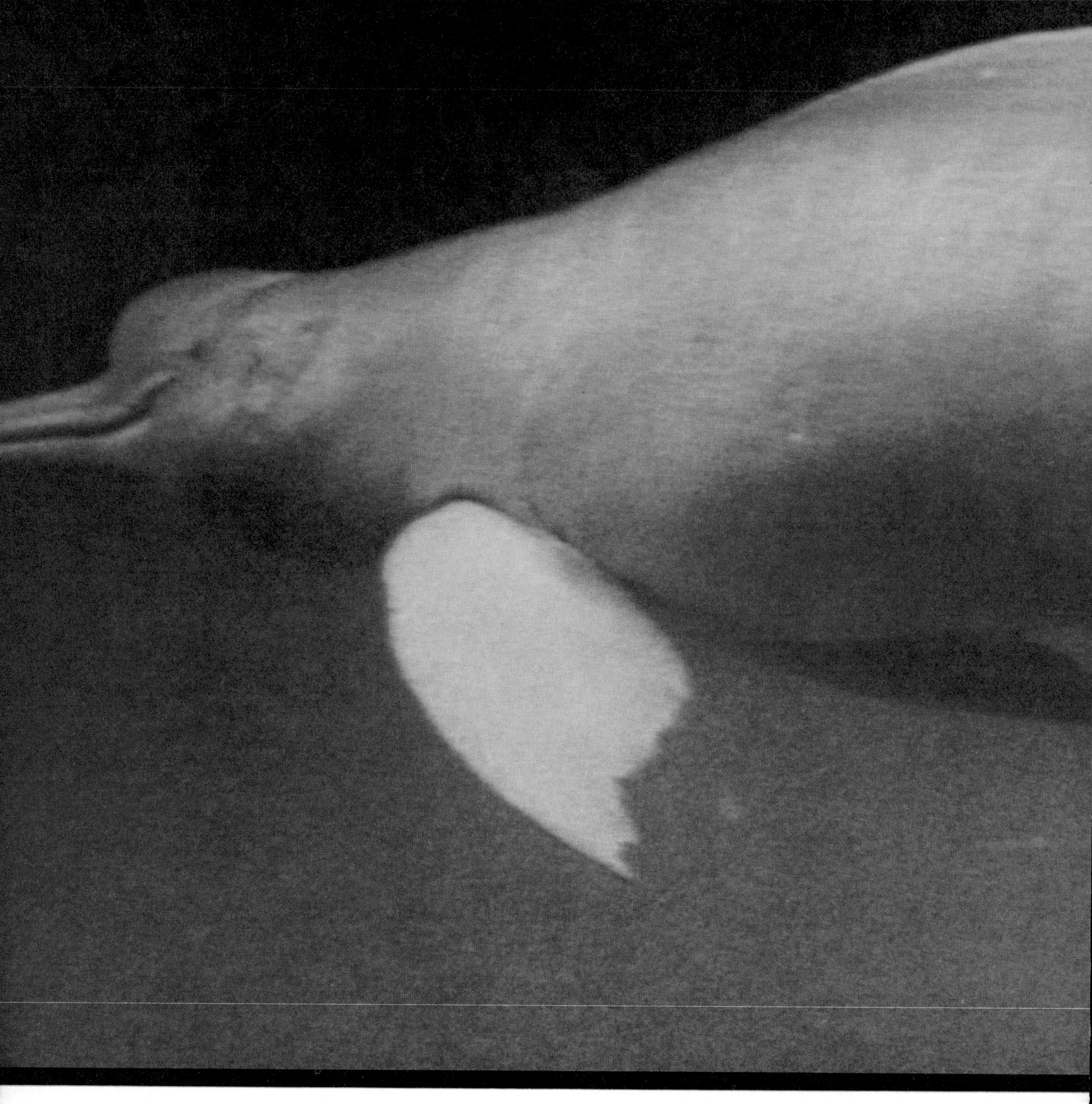

Nourishing the Young

Suckling whale calves feed on something akin to a fish-flavored milkshake. The big difference between whale milk and that of humans is the fat content. In humans it may be 2 or 3 percent. Cow's milk runs from 4 to 6 percent. But a female whale produces up to 700 quarts a day of milk with a fat content close to 50 percent. Her milk is also richer in protein, calcium, phosphorus, and thiamine. But it has only half as much water, almost no sugar, and less vitamin A than a human mother gives her baby. Whale and dolphin milk comes out thick and cream-colored to the calf suckling underwater.

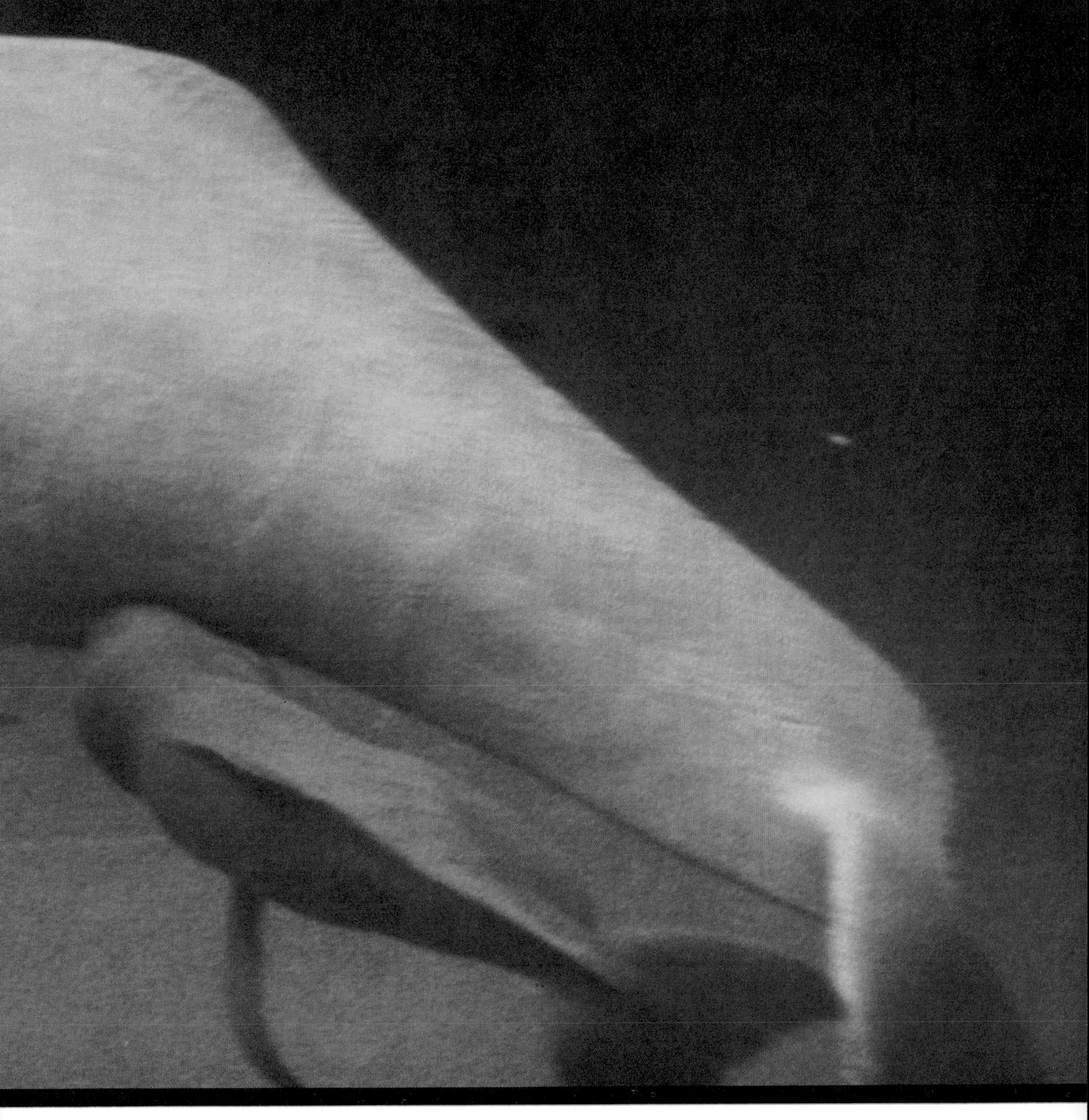

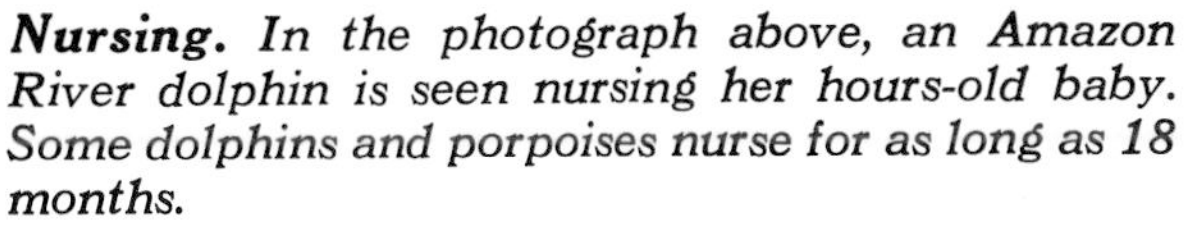

Nursing. *In the photograph above, an Amazon River dolphin is seen nursing her hours-old baby. Some dolphins and porpoises nurse for as long as 18 months.*

When the calf is very young it can't remain underwater for more than a minute at a time. Consequently it must take nourishment quickly and in short bursts. Because of these necessarily brief but frequent feedings, the milk must be rich. Usually the calf suckles by taking the female's nipple between tongue and roof of mouth. But since all this takes place while the pair is cruising slowly along just beneath the surface, the calf needs help. The female provides it by squirting milk to the calf.

Chapter V. Propulsion Needs

Water is heavier and thicker than air and moving through it requires a considerable amount of energy. But water's density provides buoyancy that is advantageous to animals that live in the sea. Very little effort is required to fight the force of gravity and all the muscular energy expended by a fish with a gas bladder can be used for propulsion—mainly forward propulsion.

Water also has greater cooling power than air. Cold-blooded fishes match the temperature of their surrounding water and spend almost no calories to make up for heat loss. On the contrary, the dolphin and other sea mammals do require a great deal of food to fuel their body temperatures. Although he is about the same size as a man, a dolphin has 18 percent of its body weight in blood compared to 7 percent in the case of man, and the dolphin's blood has more hemoglobin, which means higher oxygen-storing capacity. Thus, the dolphin is able to maintain its higher metabolism and higher body temperature.

"The tuna is an exceptionally streamlined fish, cruising at an average speed of five to eight miles an hour, but capable of accelerated bursts, for a few seconds, at up to 45 miles an hour."

The tuna is an exceptionally streamlined fish, cruising at an average speed of five to eight miles an hour, but capable of accelerated bursts, for a few seconds, at speeds up to 45 miles an hour. This animal never stops swimming. No bony fish migrate more widely through the open ocean. Fish-tagging experiments have turned up a bluefin tuna that traveled from the waters of Florida to the shores of Norway—over 5000 miles—in approximately 50 days. The tuna's body is perfectly adapted for speed and distance swimming. Their bodies are rather rigid and their tail fins are not as flexible as those of other fish. Oscillating with vigorous side-to-side strokes, these strong tails produce a powerful driving force.

Like sharks that are heavier than water, tunas have no gas bladder to keep them afloat and must keep swimming in order not to sink. The constant flexing of their muscles produces body heat 4° to 18° Fahrenheit warmer than the surrounding water. As is also true of warm-blooded animals, tunas have a higher metabolic rate of oxygen demand than most other cold-blooded fish. They swim with their mouths open so that oxygen-laden water can pass across their gills. The high metabolic rate necessitates consumption of large amounts of food. Captive skipjack tunas have been known to consume one-tenth of their body weight daily.

Sedentary animals like the sea anemone, the urchin, and the corals have little or no propulsive needs, so they can devote almost their entire food budget to growth and reproduction. On the contrary, open-sea predators that devote most of their time to the capture of food must use it for propulsion. Large fish like the marlin, sailfish, swordfish, and tuna have bodies that are three-quarters propulsive muscles and use up about 45 percent of their calories to move.

Schooling, feeding tuna. *Because tuna use so much of their energy to swim, they must search for food at all times. Their food needs are as great as those of warm-blooded sea mammals.*

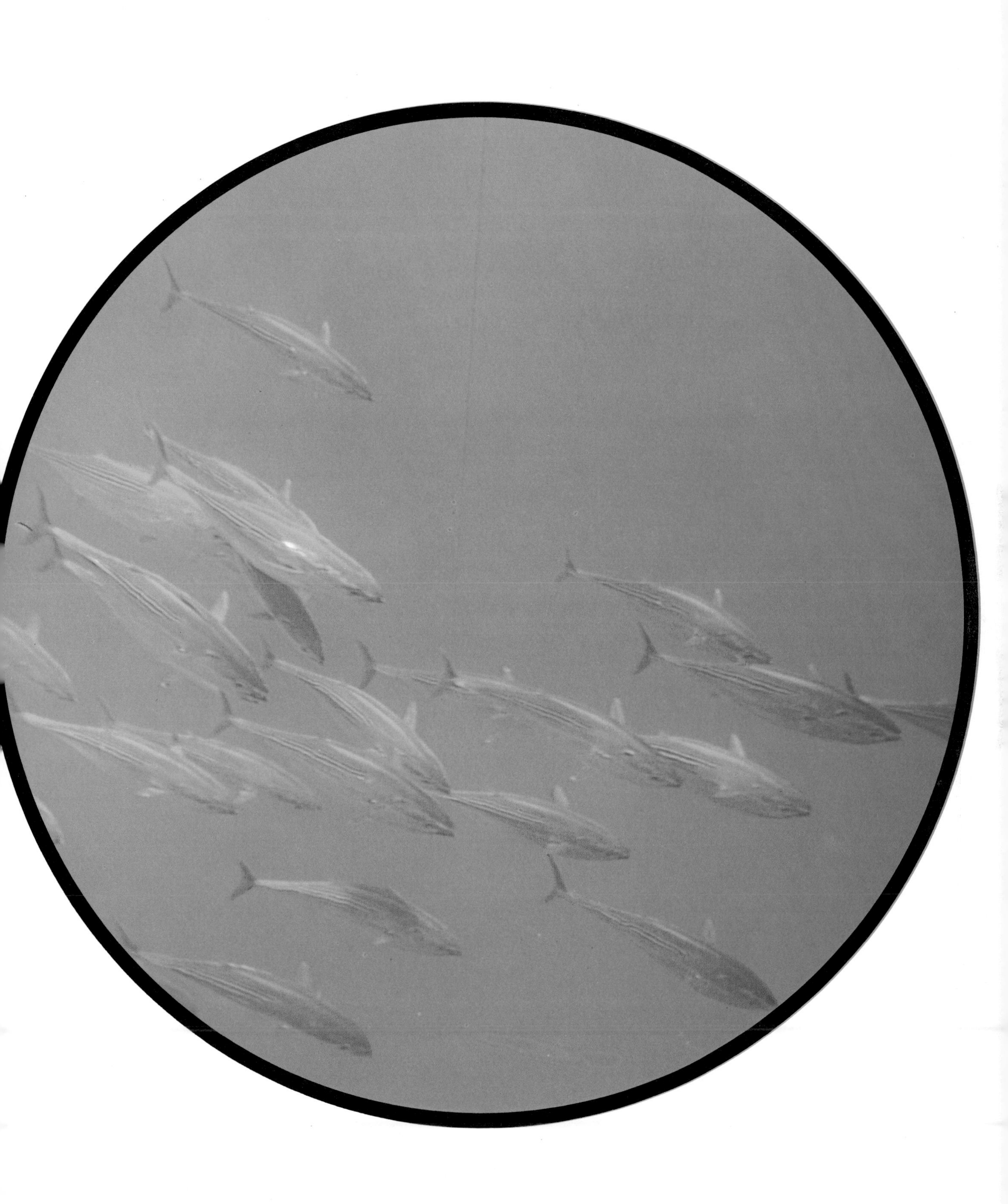

Speed Records in the Animal World

Anyone who has ever swum underwater even a short distance will spot the extraordinary statistics on this table. Water is 800 times as dense as air. For a man, trying to make speed underwater can be like a scene out of a Hollywood nightmare-sequence: the dreamer running and running but getting nowhere. Indeed, the fastest human swimmers, like Mark Spitz, do not swim "in" water but on its interface with air, in a sense "climbing" the surface with their arms while their legs and feet are pushing it behind. Still, tuna and dolphin tear through this heavy medium faster than a horse or a greyhound can gallop through air on level ground. How? Primarily because the buoyancy of water frees the animal of the demands of gravity: the entire musculature can be devoted to propulsion. Also, these are the creatures who may eat 10 percent or more of their body weight every day. Most of the energy coming in from this diet is burned up in the day's travels.

Animal	*Approximate Top Speed*
Peregrine Falcon, diving	*175 mph*
Cheetah	*72 mph*
Springbok, in bursts	*60 mph*
Grant's Gazelle	*50 mph*
Lion	*50 mph*
Tuna	*45 mph*
Dolphin	*45 mph*
Horse	*44 mph*
Greyhound	*43 mph*
Mourning Dove	*40 mph*
Spotted Hyena	*40 mph*
Cape Buffalo	*35 mph*
Giraffe	*33 mph*
Wart Hog	*30 mph*
Black Rhinoceros	*28 mph*
African Elephant	*25 mph*
Man	*20 mph*
Crabeater Seal	*16 mph*
Camel	*10 mph*
Falkland Island Sea Lion	*5 mph*
Black Snake	*3 mph*
Starfish	*30 feet/hour*
Snail	*3 feet/hour*

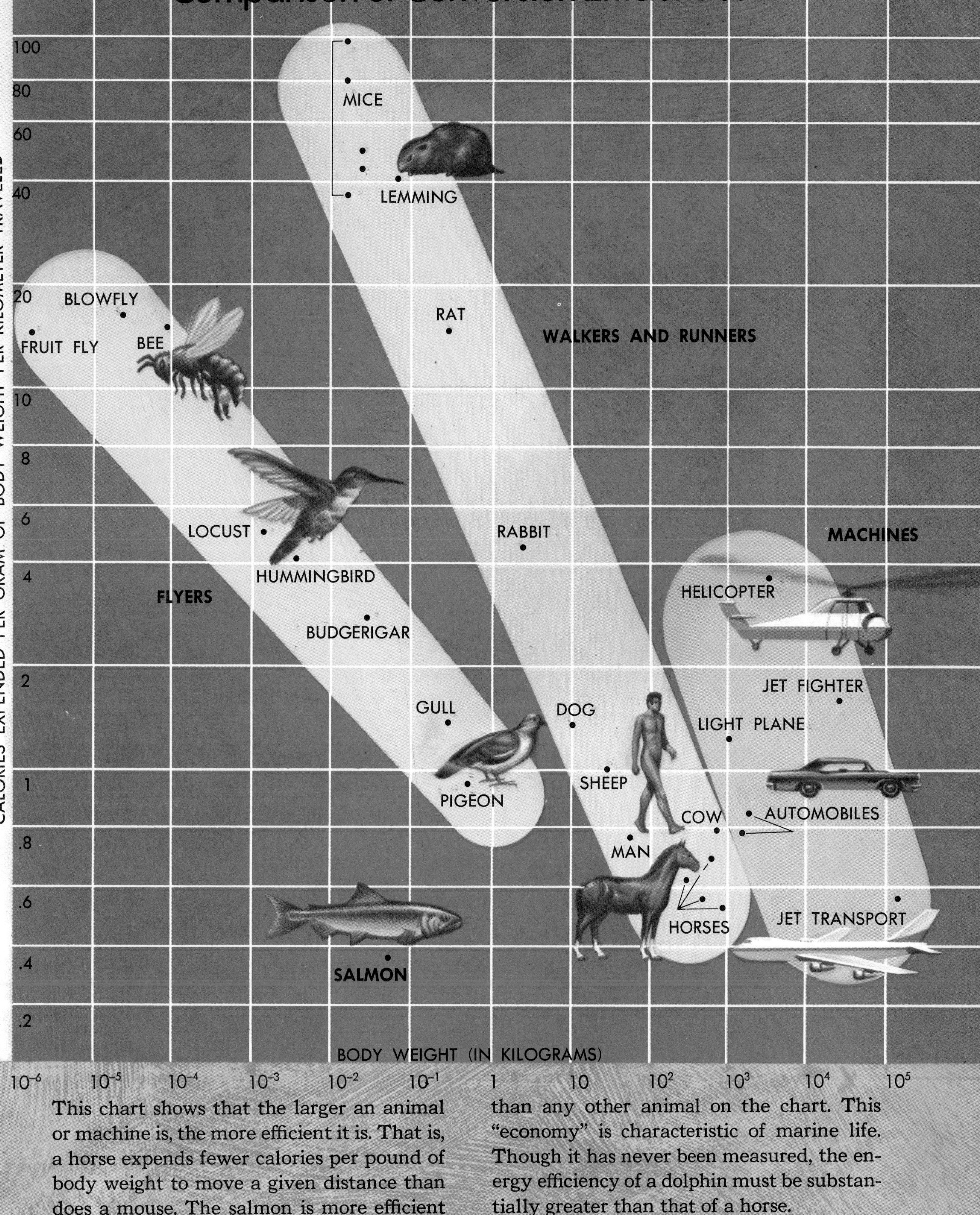

This chart shows that the larger an animal or machine is, the more efficient it is. That is, a horse expends fewer calories per pound of body weight to move a given distance than does a mouse. The salmon is more efficient than any other animal on the chart. This "economy" is characteristic of marine life. Though it has never been measured, the energy efficiency of a dolphin must be substantially greater than that of a horse.

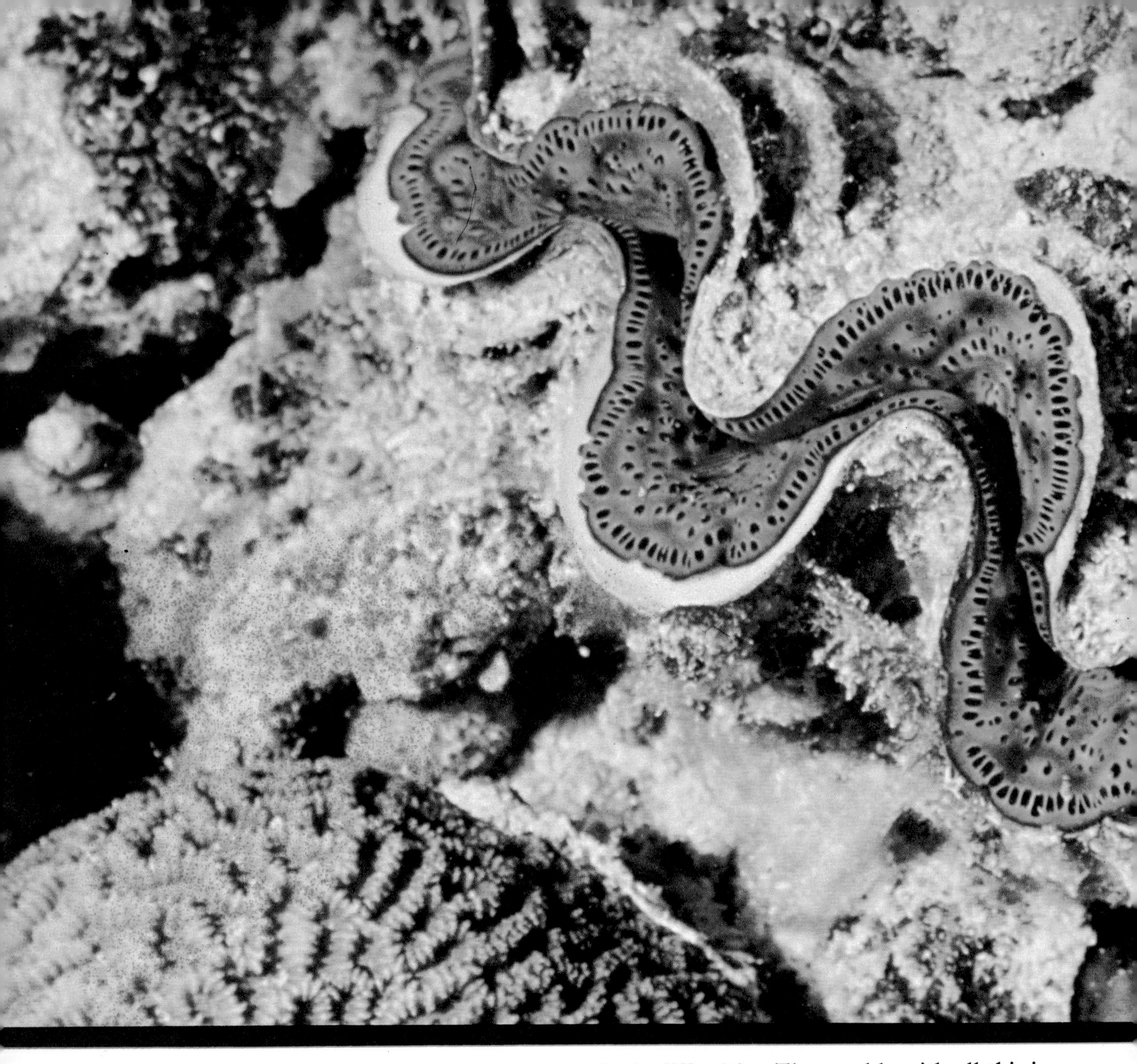

Stationary Giant

The tridacna clam of the South Pacific is the "giant bear's paw" of fearsome legend. It can grow to a weight of more than half a ton. As it tends to lie in quiet shallow water, it would theoretically be possible, I suppose, for an unwary man to step into one or, even more idiotically, to thrust his arm into the central cavity. Then, theoretically, the clam could close on the man's limb and he would be in difficulties. The trouble with all this is that I'm not sure it has ever happened. The clam is not a fast mover; it takes it a while to shut its doors. Also, the human swimmers in tridacna's habitat are not likely to clump around insouciantly on their native coral reefs. In fact, the tridacna is important to Pacific islanders. The clam's immobility makes it vulnerable to men who prize the adductor muscle as a delicacy and its shell for axes and knives.

Harmless tridacna. The thought that these animals are man-eaters is somewhat foolish. They are huge, but their shells close so slowly that they are actually almost incapable of capturing anything, and live mostly off the algae which grows inside them.

When you look closely at the mantle of a living tridacna—that purplish fringe on

"The tridacna clam of the South Pacific can grow to a weight of more than half a ton."

the open edge of the animal's shell—you see dozens of brilliant little speckles. These are tiny lenses that allow sunlight to penetrate deep within the clam's body tissues to provide light for single-celled plants that live there. Thus receiving sunlight, the algae grow and through photosynthesis produce food which is transported to other tissues by the host's white blood cells.

Chapter VI. Consumption and Conversion

To exist, most animals need three things: water, oxygen, and food. Once consumed, the three work together to maintain the body. But they must first be converted into chemicals the body can utilize.

Oxygen is converted as it acts on other materials — through oxidation — to produce energy. Water helps by carrying off wastes and by combining with the other chemicals. Food is the fuel the others help to convert. It breaks down into vitamins, minerals, fats, carbohydrates, and proteins. The oxidation of these chemical constituents is the metabolic process. Some metabolites build tissue for growth, some for replacement of old tissue or for healing sick or injured tissue in the animal body. The efficiency with which animals utilize food is a variable factor. Herbivores, for example, though they have longer digestive tracts, are less efficient than carnivores. Even the meat eaters aren't all that efficient.

"The ice on which seals emerge in the spring can be several degrees colder than seawater. Yet when seals have lain on ice, even for long periods, there are no signs of melting, which means that their skins are about the same temperature."

The means of consumption, what is being consumed, how it is consumed, and even the health of an individual animal are all factors in how efficiently food is converted. Naturally, the nature and quality of the food is an important factor too. Some animals digest food outside their bodies and take in only what they can use. The starfish does this in consuming shellfish. Some take in not only food but extraneous matter, digesting the food and rejecting the rest. The goosefish may inhale an entire lobster along with rocks and sand.

Interestingly enough, the thermal challenge to marine mammals, like the great sea lion, is not so much to protect themselves from the extreme cold of their arctic or antarctic habitats as to maintain a regulatory system which can adjust to very wide variations in temperature. In other words, if there were no more to it than wrapping their bodies in blubber the animals would overheat and die in any but the coldest conditions. Cold is of course the main problem. The ice on which seals emerge in the spring can be several degrees colder than seawater. Yet when seals have lain on ice, even for long periods, there are no signs of melting, which means that their skins are about the same temperature. How does the animal maintain blood circulation to the peripheral areas without a loss of heat that would show up by melting ice? Probably by means of an internal heat-transfer system whereby the warm arterial blood gives up almost all its heat to adjacent veins before reaching the outermost skin layers. Something of the sort, in reverse, must account for the animal's ability to bask in the sun during breeding periods without suffering heat prostration. The fats of which seal blubber is composed help too. They are low-melting-point fats, derived from the marine plankton—able to facilitate the response to sudden, drastic changes in temperature which are a feature of the marine mammal's life-style.

Landed sea lions. *These animals spend long periods of time on land. As a result they have adapted to lives both in and out of the water. Here these huge animals are sunning themselves on rocks off the coast of South Africa.*

Pastures of Stone

We have already met the parrotfish on page 5. Feeding and digestion are no simple processes for him. He needs special equipment for these chores. His sharp parrotlike beak enables him to scrape algae off coral rocks and to bite chunks off the rock in search of more. He actually grazes upon stone pastures. Special teeth in his throat pulverize these chunks of rock, yielding minute bits of algae and the tiny polyps which are the coral animal. Piles of coral debris and tooth marks on the reefs are evidence that parrotfish have been feeding in an area. For after biting the rock, and pulverizing it, the parrotfish digests the food and passes the sand. Some parrotfish grow to weigh 60 pounds. Because they extract so little food with each bite, they feed almost constantly during daytime and ingest about half their weight of material daily. One large parrotfish can extract two or three tons of sand a year from coral structures.

Probing and Cleaning

This species of sea cucumber is a shapeless animal, but valued as a gourmet treat over much of the Pacific basin. It is an enchinoderm, like the starfish, but has to some extent departed from the radial pattern common to that phylum through its habit of lying over on one side. Its tentacles probing the bottom layers for anything in the way of food particles it may find there, this sea cucumber can "process" as much as 200 pounds of sand a year. When attacked, it has the extraordinary ability to turn out its insides, enmeshing an offending crab or fish in masses of its own viscera while making its way to safety. The insides are regenerated in about six weeks. Other varieties of sea cucumber feed by extending sticky, many-branched tentacles out to catch passing plankton. As the tentacles become full, the sea cucumber puts a food-laden member in its mouth and sucks it clean, then returns it to catch more food.

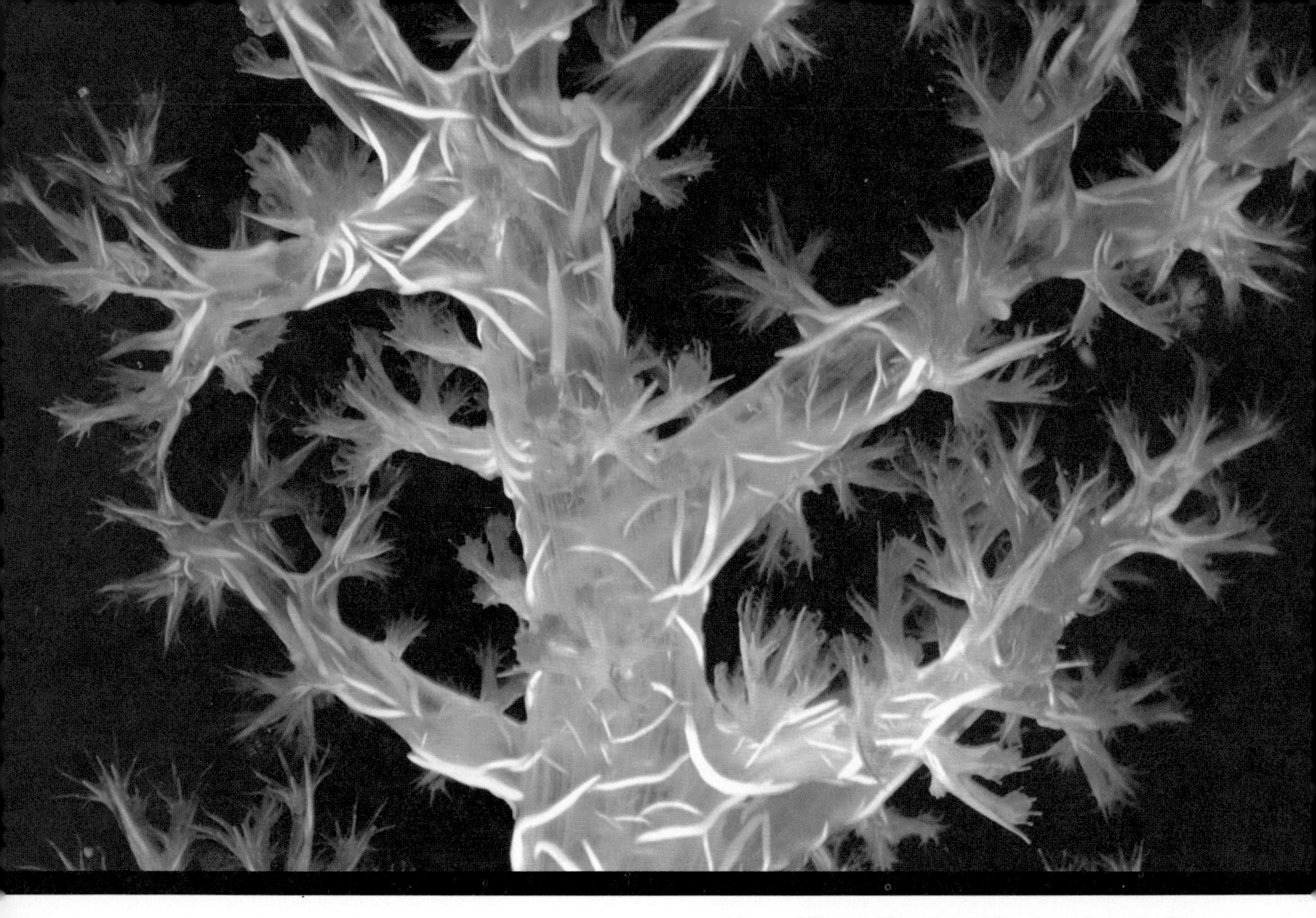

Gentle Collectors

Like the flowers they resemble, soft coral communities wave gently in the surge of the sea, collecting passing microscopic crustaceans. These are alcyonarians, members of a diverse order that includes the organ-pipe coral of the tropics, gorgonians like sea whips, sea fans, and sea plumes that grow at all depths in the ocean, sea pens and sea pansies that attach themselves to soft muddy bottoms. Sea fans are popular collectors' items, and red-and-black coral is highly prized as jewelry and has become very rare. Unlike reef-building corals, alcyonarians have only spicules, or needlelike parts of hard calcium carbonate, for their internal skeletons, which fall on the sea floor when they die, adding to the sandy bottom. Soft corals are made up of colonies of individual polyps living in close harmony with each other. They live mostly in the shallows where sunlight nourishes the minute algae that are the food of tiny crustaceans. With minute hairlike cilia, the alcyonarian corals set up currents of water to drag the crustaceans into their mouths. The feeding polyps of alcyonarians resemble those of true hard corals and function in the same way. Eight tiny tentacles surround the mouth and capture plankton. Before settling into their stationary adult stage, the larval coral animals are free-swimming individuals, living separately and feeding on the sunlight-fed algae and the zooplankton.

"With minute hairlike cilia, soft coral communities set up currents of water to drag the crustaceans into their mouths."

External Digestion

Not having teeth, starfish have evolved a unique method to digest food. Depending upon touch and smell to locate a prey, their success in locating a meal is related to the amount of territory they can cover. The undersides of their arms contain hundreds of slender feet, always moving, armed with tiny suction discs. Movement is achieved by gripping a rock or any object with the feet, retracting them—thereby creeping slowly along. In areas where food is plentiful, starfish move at a speed of less than 15 to 30 feet per hour. In the deep sea, where food is less abundant and more ground must be covered, it is believed some species move at the speed of 100 feet per hour by running on the edges of their arms rather than dragging with their feet. Their diet consists mainly of shellfish and bottom detritus and debris.

When a meal, perhaps an oyster, is found, the starfish engulfs it, attaches its many suction-cup feet to the sides of the shell, and secretes its digestive enzymes. Even though the shells are tightly shut, the mollusc will

"As the muscles of the bivalve relax, the starfish opens the shells and everts its stomach to digest the meat inside."

eventually feel the effects of the chemicals. As the muscles of the bivalve relax, the starfish opens the shells and everts its stomach to digest the meat inside. The enzymes reduce the meat to a fluid which can then be absorbed by the walls of the stomach. After digestion, the stomach is withdrawn. One species of starfish was observed eating 50 small clams in six days.

The Food Chain

The food chain starts with that primal source of energy, the sun. That part of the solar energy falling on the ocean which remains available to ocean life is due to ocean plants in the plankton, the "primary producers." Tiny herbivores eat these plants, tiny primary carnivores eat the herbivores, larger secondary carnivores may eat both primary carnivores and herbivores, and so on. Each group is called a trophic level. And at each successive remove from the sun there is a substantial loss of the original solar energy.

At first it was thought that the average efficiency of each trophic level beyond primary producers was about 10 percent. But a few

> "Two billion calories of the sun's energy arrive on a square kilometer of the ocean's surface each day of the summer. About 99.5 percent of this energy is reflected, scattered, and absorbed by the ocean, while only about 0.5 percent is used by phytoplankton."

years ago we accumulated data suggesting that food-chain efficiencies may be closer to 20 percent. Let us see how it works. Imagine the transfer of matter and energy through a typical food chain in the ocean off southern California. Two billion calories of the sun's energy arrives on a square kilometer of the ocean surface each day of the summer. About 99.5 percent of this energy is reflected, scattered, and absorbed by the ocean, while only about 0.5 percent is used by phytoplankton to produce 1,670,000 grams (3682 lbs.) of carbon. The phytoplankton respire 32 percent of this as carbon dioxide and excrete about 8 percent as dissolved organic matter, so that 1 million grams (2205 lbs.) are available for grazing by herbivores. The main herbivores are flea-sized crustaceans and microzooplankton (protozoans and young forms of zooplankton). Together these organisms eat 93 percent of the phytoplankton, leaving about 7 percent to be degraded by bacteria or to sink to the bottom to be filtered out of the water and eaten by bivalves and worms. The herbivorous zooplankton egest 20 percent, respire 42 percent, excrete 3 percent, and give off 5 percent as molted exoskeletons. Fifteen percent of the food goes toward growth and the final 10 percent toward reproduction. Thus, 232,000 grams (511.5 lbs.) of carbon per square kilometer is produced by herbivores, and their efficiency is 23 percent.

The carnivorous zooplankton—some copepods, small jellyfish, arrow worms, larval fish—eat about 60 percent or 140,000 grams (308 lbs.) of the herbivorous zooplankton. The remaining 92,500 grams (204 lbs.) go to the bacteria and bottom-living life except during some times of the year when the carnivores eat almost all the herbivores. The average carnivore is quite efficient. It ingests 10 percent, excretes about 5 percent, does not molt in most cases, and respires about 40 percent. Growth and reproduction account for 45 percent of the ingested food. The food chain efficiency at this trophic level is about 27 percent.

It is estimated that anchovies eat about 50 percent of the accessible food, so it is probably a safe guess that all other small fish eat another 25 percent. A study of sardines showed that the small fish are rather inefficient transferers of energy. They egest 10 percent and probably excrete little, but they are such active swimmers that 80 percent of their energy is used for respiration.

The Pyramid of Life

In the sea approximately 10,000 pounds of plants will support 1000 pounds of plant-eating animals, which in turn will support 100 pounds of meat-eating animals, which in turn will support 10 pounds of tuna, which will support one pound of human flesh.

Decelerating Discs of the Food Cycle

We may imagine the food cycle as a series of eccentric discs spinning at differing speeds around a revolving center. At this center phytoplankton marries the energy of the sun to nutrients in seawater to produce organic material. The turnover in the central core is rapid—the lifetime of each of these myriad, miniscule planktonic plants is short.

One of the eccentric discs carries zooplankton, masses of eggs, larval forms, and small animals. Turnover time here is short too. Growth of the little animals is rapid. If they survive, they move to the next disc in a short time.

Many don't make it, so the third disc is not so densely populated, nor does it move so fast. The animals at this level are adults of small species and juveniles of larger fish. Some will spawn and complete their lifespans in this level. For them there will be no

next stage. Some will be lost to larger predators. Flesh is more expensive on this disc; every pound gained here represents 100 pounds of phytoplankton.

The fourth, still larger disc holds fewer animals but bigger, longer-living ones. They require more space, need more food than the individuals on prior discs, and the oceans can support only a limited number of them.

Mammals and the most voracious of fishes, the sharks and tunas, populate the final disc. Spinning off each of the discs is a steady stream of organic material, dead tissue, and detritus. Bacteria in seawater and on the ocean's floor work to liberate the constituent chemicals in this material.

> "At the center of the food cycle phytoplankton marries the energy of the sun to nutrients in seawater to produce organic material."

Chapter VII. Locating a Meal

The difficulties involved in locating a meal in the ocean are partly determined by two major respects in which the sea differs from the land. First, the sea is a three-dimensional space rather than a two-dimensional one. Second, water at its most transparent is not nearly as clear as air. Sight is most often the sense by which a meal is captured, but the initial location of a potential prey is often through other senses that are not limited by water clarity.

Because of the rapidly increasing pressure (14.7 pounds per square inch for every 33 feet) and the rapid falloff in sunlight (there is little left below 1000 feet), most sea animals cannot range as freely in the vertical dimension as can, for example, birds (or, recently, man) in the air. Still, there are animals specially adapted for "breaking the pressure barriers." The sperm whale can dive down about a mile in search of the giant squid, which for its part probably never voluntarily comes anywhere near the surface (and a living specimen of which has so far as we know never been caught or even seen by man). Other marine mammals—like the sea lion, seal, and dolphin—can dive perhaps a quarter as deep.

A significant problem for active feeders is the poor visibility underwater. In navigating through their three-dimensional water space in search of their meals, fishes employ what we call their "sixth sense": the lateral line organ—a longitudinal system of canals and openings which starts over the head of most fishes and runs under the skin along the body to the tail. This system almost certainly has two jobs: that of perceiving distant pressure disturbances as well as changes of flow patterns close to the fish. It is a sort of "auditory ear." As a fish swims forward, he pushes ahead of him a small pressure wave. The strokes of his fins also generate pressure waves. So a creature equipped with a lateral line organ is aware of a world of invisible and inaudible messages informing him about what other creatures do and where. This additional sense helps a predator locate a meal, but at the same time warns the prey of the predator's approach.

Marine mammals have developed "echolocation"—a sonar apparatus like the bat's, capable in the dolphin and other species of elaborate sophistication. Animals can also taste and smell the water. Sharks can home in on ounces of blood diffused in the sea.

"Because of the rapidly increasing pressure and the rapid falloff in sunlight, most sea animals cannot range as freely in the vertical dimension."

Like a human fisherman, the sea creature in search of a meal has to know where to go. Large regions of the ocean are a desert so far as life is concerned. Animals must conserve their energies by concentrating their hunting in those areas of the sea where instinct tells them food will be found; otherwise they will be lost in the wastes, and they will die of starvation. So close is this affinity for the ocean's dynamics that many sea animals need spend only a few moments in the mornings and evenings to earn themselves an ample diet.

Gray angelfish. *These fish usually travel alone or in pairs and are found around coral reefs in tropical waters. They feed on invertebrates like crabs and barnacles.*

Dolphin Sonar

As we saw in *Oasis in Space* the dolphin is an acoustic animal. On the "clearest day" in the sea the eye cannot reach much farther than 100 feet. Even crystal-clear water is a fog so far as the organs of sight are concerned. So, in returning to the sea, the dolphin's ancestors adapted to these limitations. How to locate a meal in a fog? He compensates by means of the hypertrophy of his auditory faculty; his internal and external ears are greatly modified in structure in order to receive and interpret a wide range of water vibrations. All this not only vastly improves the animal's hearing underwater but refines it for the purposes of echolocation, that sonarlike process with which the dolphin homes in on objects as distant as a mile away that he wants to inspect—perhaps to eat. The dolphin is not the only animal in the sea who echolocates; most of his whale cousins can do it, also such other mammals as the California sea lion.

In practicing echolocation, the dolphin emits a fascinating spectrum of noises—clicks, burst-pulse signals, pure tones. The modulated sounds are for the purposes of socializing—remember, an activity almost as important to the dolphin as eating. Others—clicks lasting from a fraction of a millisecond to as long as 25 milliseconds—are for echolocation purposes: low-frequency "orientation" clicks to give the animal a general idea of his situation; high-frequency "discrimination" clicks to give him a precise picture of a particular object he is interested in.

The dolphin's larynx is an intricate structure, without vocal chords, but provided with a sphincter enabling the animal to

"In practicing echolocation the dolphin emits a fascinating spectrum of noises—clicks, burst-pulse signals, pure tones. The modulated sounds are for the purpose of socializing, an activity almost as important to the dolphin as eating."

make his sounds at the same time he is feeding underwater. Many experiments have shown us how finely tuned the apparatus is. At ten yards or more a dolphin can discriminate between two objects only a few centimeters in diameter. He can pass through a complicated obstacle course of thin wires strung between himself and his objective.

Importance of sound. *Humans and dolphins can see equal distances underwater. It is the dolphin's sonar that gives him a clear advantage over man—in locating prey, predator, playmates.*

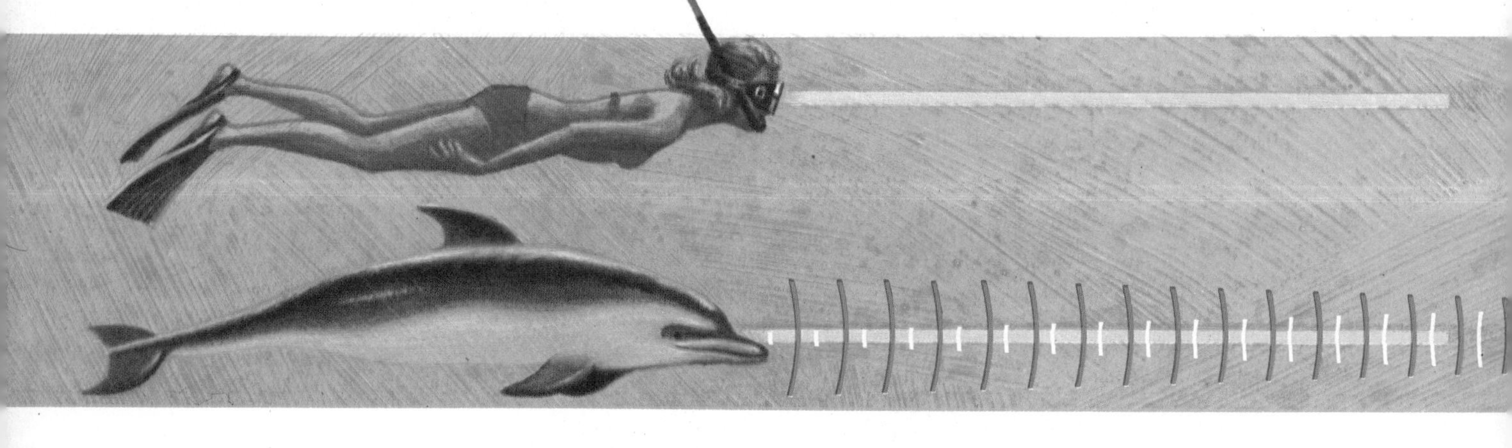

Wide-Angle Eyes

Among many fishes and other marine creatures sight takes over where other senses leave off. Or it may augment the sense of smell or the pressure-sensing lateral line system. So finding food may be a combination of any two or all three senses.

> "The eyes of the hammerhead shark aren't so unusual. But their placement is. They stand at the ends of freakish heads, reminding one of a Picasso painting."

Smell may attract the attention of a fish to a potential meal. Then perhaps the sensation of pressure is felt by the "lateral line." But as the fish closes on its prey, sight is often necessary for the final coup. This holds true for other animals of the sea besides fishes. The octopus and squid, both molluscs, have highly complex eyes that compare favorably with our own in visual acuity. The eyes of the hammerhead shark aren't so unusual. But their placement is. Their eyes stand at the ends of their freakishly shaped heads, reminding one of the strangely placed eyes of a Picasso painting. This unusual placement may give the hammerhead the widest possible view of the sea around him. It also may serve to separate the nostrils, giving him a greater ability to determine the direction of a scent corridor; in other words, to help him find the source of an odor.

The unfishlike fish called the seahorse is one of the several varieties that has binocular vision. And many of the sea basses have the same faculty—combining two pictures into a single image in depth.

The Big Eyes

Unlike land animals, this octopus and other marine animals have spherical lenses to focus the light rays for clear underwater vision. Light rays are bent upon traveling from a medium of one optical density to a different density. Because the optical properties of water are very much like those of the eye, the light rays are bent very little in passing from water through the cornea and lens. Consequently a lens of relatively great curvature is required to bend or focus the light adequately. Animals that inhabit dark caves or great depths have large eyes to intercept more light and focus brighter images on the retina. Below 1000 feet, where a special deep-sea fauna spends most of its life, big-eyed fishes must be able to catch the occasional flashes of luminescence from passing organisms. In some cases, these fish are especially sensitive to the faint wash of blue light that penetrates the sea: their eyes have many blue-sensitive cells on the retina.

"Below 1000 feet, where a special deep-sea fauna spends most of its life, big-eyed fishes must be able to catch occasional flashes of luminescence."

In the seagoing reptiles—turtles, marine iguanas of the Galápagos, and the highly venomous sea snakes of the Pacific and Indian oceans—a transparent shield protects the eyes. In the lizards and turtles this shield is a third eyelid. In the sea snakes it is a rigid plate which is shed with the rest of the skin.

The mudskipper, a fish that spends time on dry land, has a fold of skin over its eyes to keep them moist.

Testing for Taste

Most of the nearly 4000 known species of catfishes have special sensory barbels, or "whiskers," extending from their lower jaws.

The striped sea catfishes, shown here, school densely and approach the ocean floor to grab around for food. The barbels serve a dual function. They locate food in the mud by touch as well as taste, to see if it is palatable. Through the use of these barbels, catfish feeding is not dependent upon sight, which is affected by water clarity.

A number of other families of fishes have barbels serving the same function as the catfishes'. Theirs too are movable extensions of the skin, rich in nerve endings for tactile and taste senses. The sturgeon, for example, cruises the ocean floor—keeping contact with four short barbels located inches ahead of the suctorial mouth on his ventral side. When the sturgeon tastes something good,

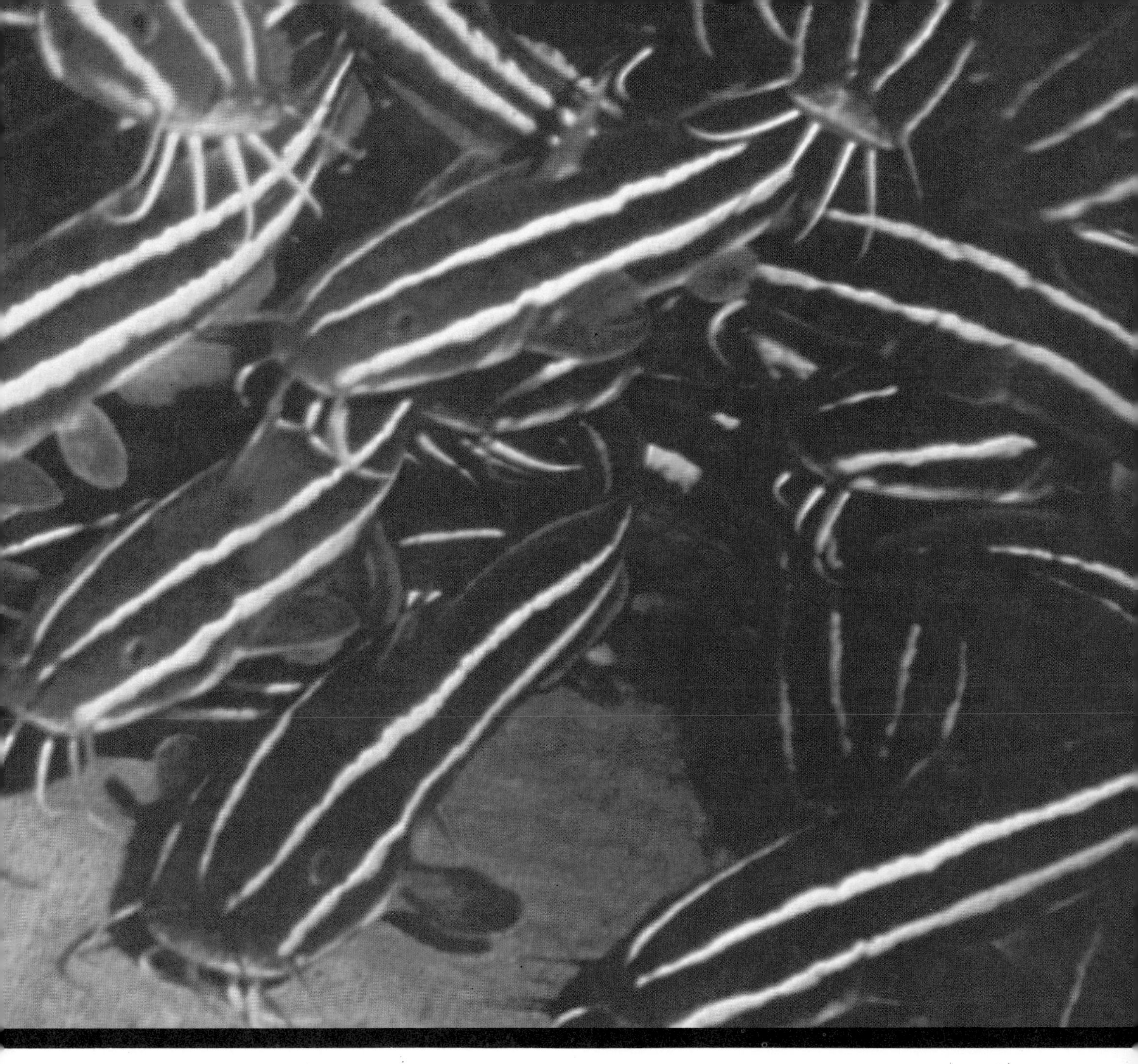

he extends his protrusible mouth and sucks up the delicacy.

> "In some deep-sea fishes the barbels—sensory whiskers—include special light-producing organs for attracting prey or mates."

Several families of deep-sea fishes carry a single barbel on the chin. In some it is long and heavily branched. In others the barbels include special light-producing organs, used to attract prey or members of its own species for mating. Another application of sensory organs similar to barbels is found in the mud hakes of the North Atlantic. These relatives of the cod have a single barbel; but their pelvic fins have also been modified into long, threadlike devices. The hake carries these highly specialized fins off to the sides. They have evolved into feel-and-taste organs.

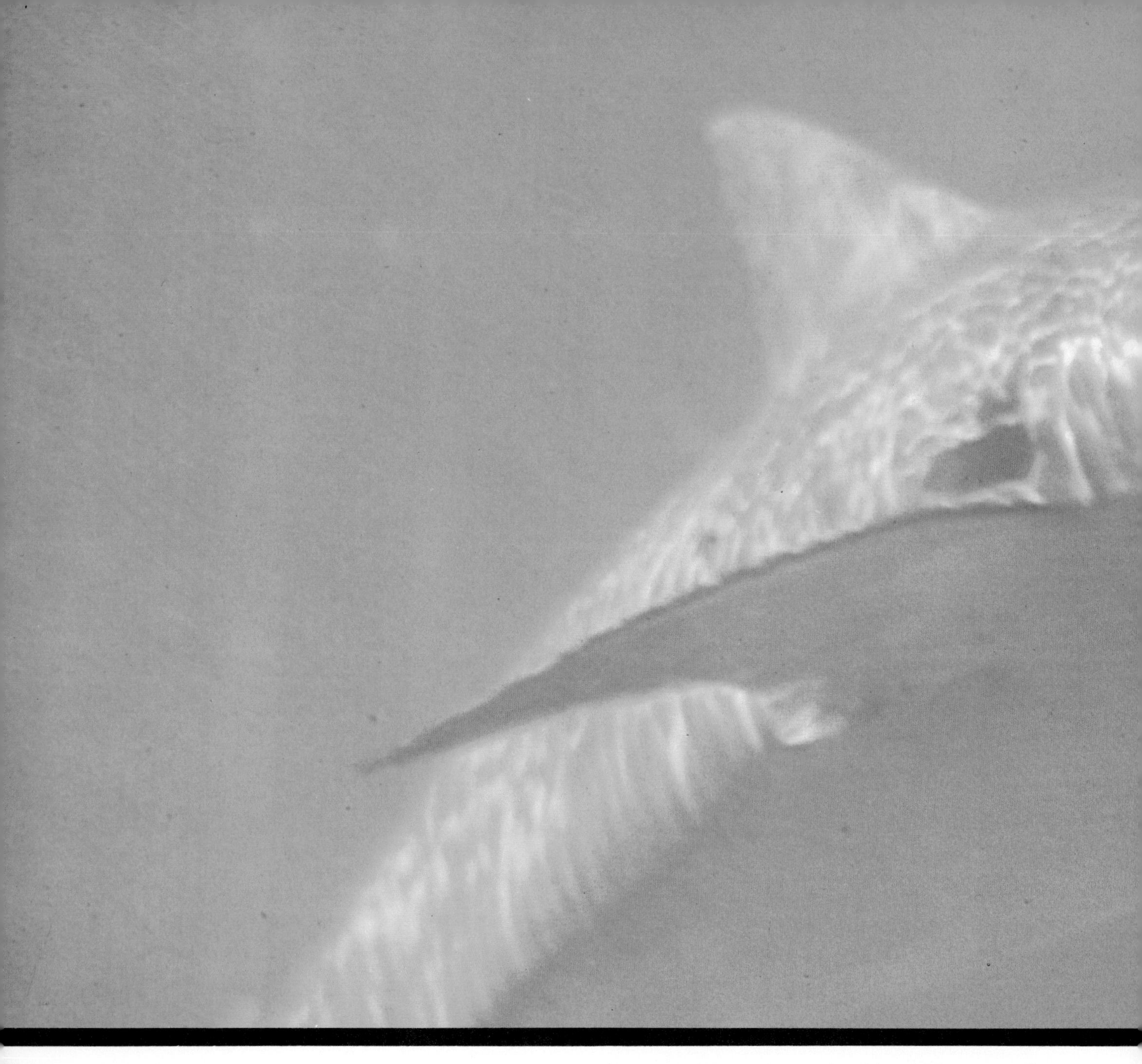

The Nose Knows

Sharks can zero in on a wounded fish from a quarter mile away or more. Night-feeding fishes can find food without light. Eyeless cavefishes can too. How? Through a particularly acute sense of smell.

The shark swims in circles or random patterns, picking up odor gradients and, poor eyesight or not, pinpointing the source of an attractive smell. The tiger shark shown here has nostrils on the underside of his snout. So do most other sharks. Other fishes have nostrils on the top or sides of their heads. In most, the nostrils are blind sacs not used for respiration but lined with odor-sensitive membranes.

"Sharks can zero in on a wounded fish from a quarter mile away or more."

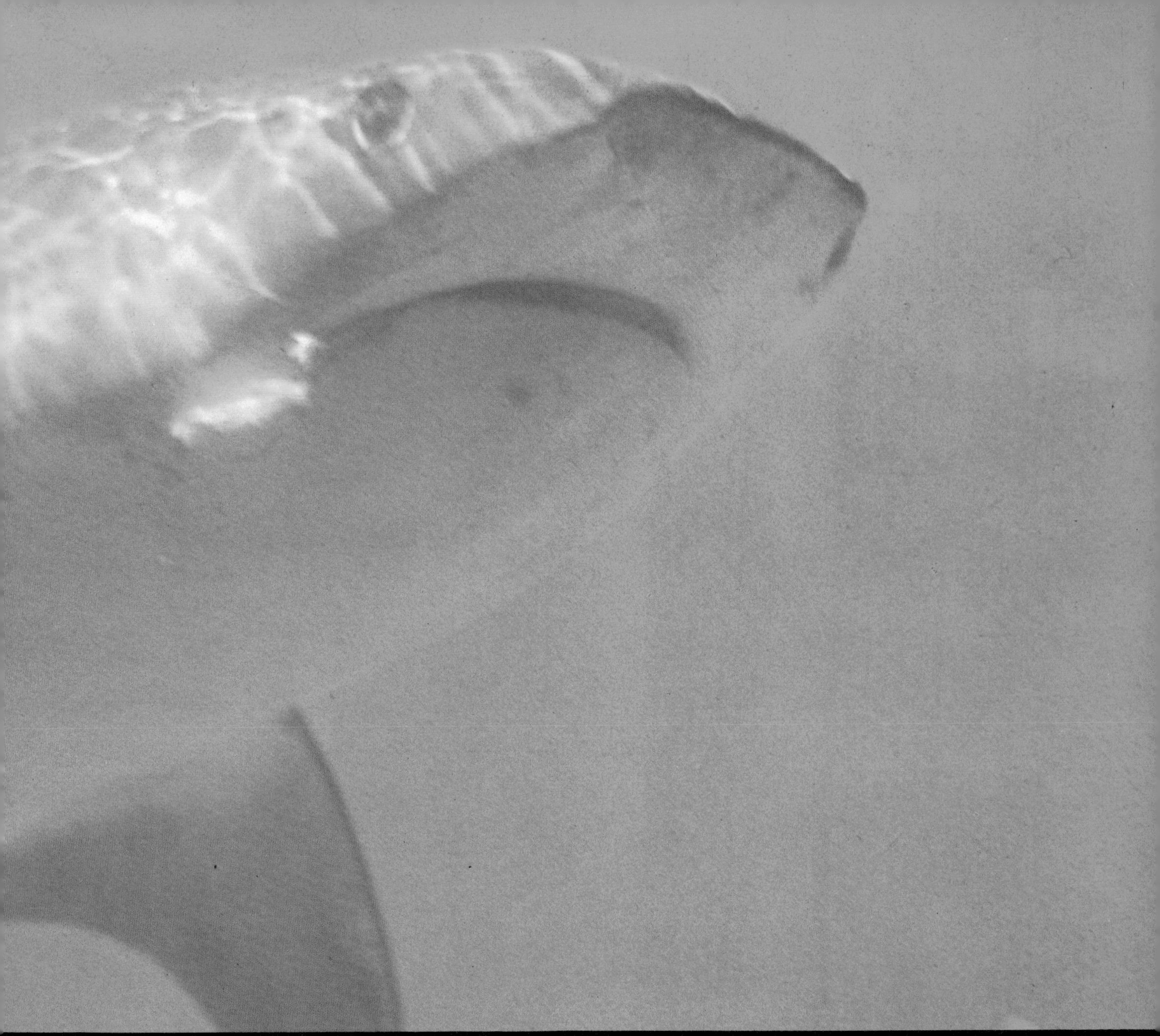

Once, aboard *Calypso,* we tested the United States Navy's shark shield—the inflatable, plastic Johnson shark screen. Although the thin plastic was no protection from the powerful thrust and slashing teeth of a shark, it prevented man's scent from entering the water. The sharks attracted to the area passed the plastic bag with little more than cursory examination. But many times we have seen sharks give more than a passing glance at our unprotected divers in the water.

Other experiments have shown catfishes, sharks, morays unable to locate food when their nostrils are plugged. Some biologists, seeking to learn the secret of how the salmon can find its native river after years at sea, believe it depends on smell. Each river has its own individual smell and taste to these fish. And some primeval force imprints this scent in the brain of the salmon in its first year of life in the fresh water. Years later, coming home to spawn, the memory of this scent helps the salmon home in.

Limbs and Bristles

The spidery-looking banded shrimp uses its head in several ways to capture food. Besides four or five pairs of legs originating on its body, the shrimp has two or three pairs of appendages emanating from its head. These are antennae, antennules, and mandibles. They serve many functions including locomotion, feeling, setting up water currents, chewing, and seizing. The body legs also help crustaceans catch food. Typically a shrimp or other crustacean might first feel its planktonic prey, then move in more closely. The appendages provide information about taste. When close enough, it can set up water currents that will bring the tiny animals to its mandibles, with which it chews the food. In some lobsters and crabs, bristles on the body limbs also have a sense of feeling.

Busy Fingers

The big-headed, bottom-dwelling sea robin appears to walk along the ocean floor with "fingers" feeling for food everywhere he goes. These "fingers" are really the first two or three spiny rays of the fish's pectoral fins. The rays are separate from each other and from the rest of the pectorals. They are busy almost constantly, sifting the sand or hard-packed mud, turning over bits of gravel or small stones, seeking marine worms, tiny crustaceans, and some of the smaller molluscs. While the fin rays drum along the bottom, the other flat, broad pectoral fin glides just off the bottom, sculling slightly to keep the fish moving over new ground in the food hunt. Several species of sea robin inhabit the shallows and moderate depths of the Atlantic coast from New England to Venezuela.

Chapter VIII. Sedentary Feeders

Many animals in the sea fasten themselves in one place and spend the whole of their lives there. They have found evolutionary success by limiting movement and conserving energy. Some affix themselves to manmade structures like ships or piers, to the bottom, to shells of other animals, or even to swimmers like whales. Some of them are able to propel themselves but do so rarely.

> **"Many animals in the sea fasten themselves in one place and spend the whole of their lives there."**

Fixed animals are dependent upon their environment to deliver food to them. They have developed remarkable methods of collecting it. Bivalves—clams, mussels, and oysters—filter their food from seawater pumped through their shells. With hairlike appendages called cilia they agitate the water, keeping it moving to bring in new food. Most of the bulbous shapes we know as sponges are the structures of colonies of individual creatures living together. The large central opening seen in some is not the mouth, for each of the individuals feeds separately. The structure supports the animals as they strain microscopic bits of food—plant and animal—from water they pump through the porous skeleton and out the large central opening.

Filter-feeding worms like the beautiful feather-duster build stiff tubes around their bodies into which they retreat at the slightest disturbance. To feed, they extend feathery arms from the tube and strain out any food particles the water carries to them. In similar fashion barnacles feed, filtering the water with modified legs to catch whatever comes by. Some barnacles use the motive power of others by attaching to animals that will carry them about the oceans.

Some so-called sedentary feeders, like mussels, move the water rather than move themselves through the water. Bivalves draw water through their gills where a mucous sheath is secreted. Waving, hairlike cilia beat water into the sticky mass. This entangled plant and animal matter flows down special grooves in the gill filaments and enters the mouth. Some species of mussels are able to circulate as many as nine gallons of water through their shells every hour. Along with some relations among the molluscs, mussels are considered the source of the Golden Fleece. For their extended foot can excrete thin threads of an extraordinary material, the byssus—threads which the mussel uses to haul himself along on or to anchor to rocks or other mussels.

A few marine creatures have adapted to living in or eating the wood of ships and pilings. Their shells are fragile, no longer needed for protection, but lobed with fine ridges bearing sharp teeth. Some borers have an enzyme to break down cellulose and make wood into a digestible food.

The propulsive methods of sea urchins, starfish, snails, and nudibranchs do not give them much chance of capturing healthy, fast-moving fish. Instead, they rely mostly on fixed or floating food for sustenance: kelp, coral polyps, shellfish, detritus, plankton, and members of their own species.

***Mussels.** There are over a dozen varieties of mussels, most of which are found near the shore. They attach themselves to pilings and rocks by means of a strong, threadlike material called byssus.*

▲A

Fixed Feeders

Half a billion years ago there were sponges on earth. Sponges feed by simply sitting and eating. They remain always in one place, moving cilia imperceptibly to bring food-laden water into their bodies, then expelling it when they've filtered out the nutrients. The corals are not as old as the sponges, but there have been corals of various kinds on earth for hundreds of millions of years. The reef-building corals and the soft corals and gorgonians, like those shown here, are almost all colonial animals, interdependent as they live and die side by side.

A / Sponge. *All varieties of sponges have collar cells with cilia which are used to wave water gently into their bodies. The plankton and nutrients are filtered and used, the rest of the water expelled.*

B / Sea pen. *This animal is even more passive than the sponges. It merely sits, fixed to the bottom, and allows the water's natural currents to pass through it. And as the water passes through, the sea pen eats.*

C / Bryozoa. *These colonial animals too are filter feeders. They remain attached to a rock or piece of wood and allow the currents to carry food to them.*

▲B ▼C

Garden eels. *These independent animals live in colonies but always fend for themselves. They sit in the burrows they've built, rarely venturing outside, and wait for currents to waft food in their direction.*

Alert blenny. *This fringehead blenny is a small fish which lives mostly near the shore. It usually constructs a tubular home in the sandy bottom but sometimes dwells in empty cans and bottles.*

Sitting and Waiting

Many animals and fish dig holes in the muddy bottom of the sea floor or take over a rocky crevice. Some just remain there waiting for something edible to approach, others occasionally venture outside.

Most blennies are bottom-dwelling carnivores or omnivores. They establish a territory around their homes and defend it vigorously. Much of their time is spent near home, patrolling their territory, searching for food. Blennies travel from place to place in short bursts. One blenny that lives around tropical reefs resembles a cleaner wrasse in the shape of its body as well as its coloration. It even imitates the dance the wrasse performs to attract customers to its cleaning station. When a customer (not expecting to be the victim of a fraud) approaches for a cleaning, the imposter gives him a series of bites.

Garden eels stretching from their holes look like cobras rising from a snake charmer's basket. The territorial eel digs a tunnel several feet deep 50 to 200 feet below the surface, claiming all the adjacent area. Each tunnel is about six feet distant from those around it. Garden eels live in colonies of 200 to 300 individuals. They are extremely cautious and withdraw into the safety of their homes at the first indication of movement around them.

▲ A

Fixed Strainers

These "fixed strainers" resemble flowers. They draw water through their bodies. But fixed strainers extend into the water, in some cases from tubes and protective calcium plates, specialized limbs once used for propulsion. The appendages that moved the animal moved the water at the same time. Because the water bore enough food, it finally wasn't necessary that the animal move. This sedentary way of life has been adopted by crinoids, feather-duster worms, and the shrimplike barnacles. The extensions on each arm or leg of the fixed strainers form a fine net to screen microscopic food particles out of the water. These are caught by the sticky mucous film coating the limbs or in the fine web of the strainer.

A / Feather-duster worm. *Fixed to a rock or other solidly moored object, these creatures filter drifting plankton from the water passing through them. They are very sensitive to light, and will withdraw in response to a passing shadow.*

B / Feather star. *The arms of this delicate crinoid are lined with cilia which stop plankton as it drifts by. Microscopic animals and plants are then transferred to the animal's central mouth by the arms.*

C / Crinoid. *Crinoids, also known as feather stars and sea lilies, are exquisitely beautiful animals which are either free-swimming or stalked. This crinoid, like its relative, above, feeds by waving its arms and gathering plankton.*

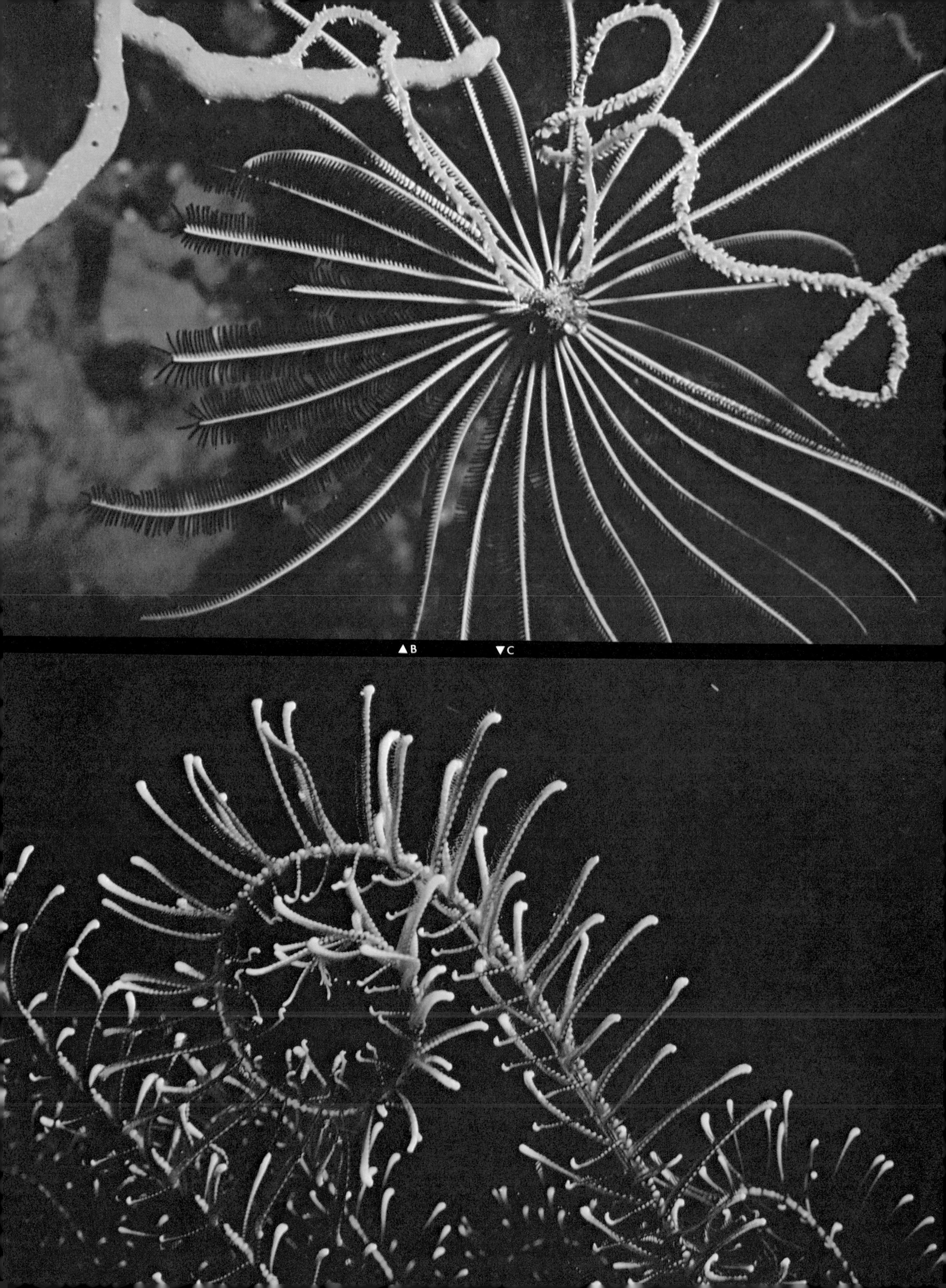
▲B
▼C

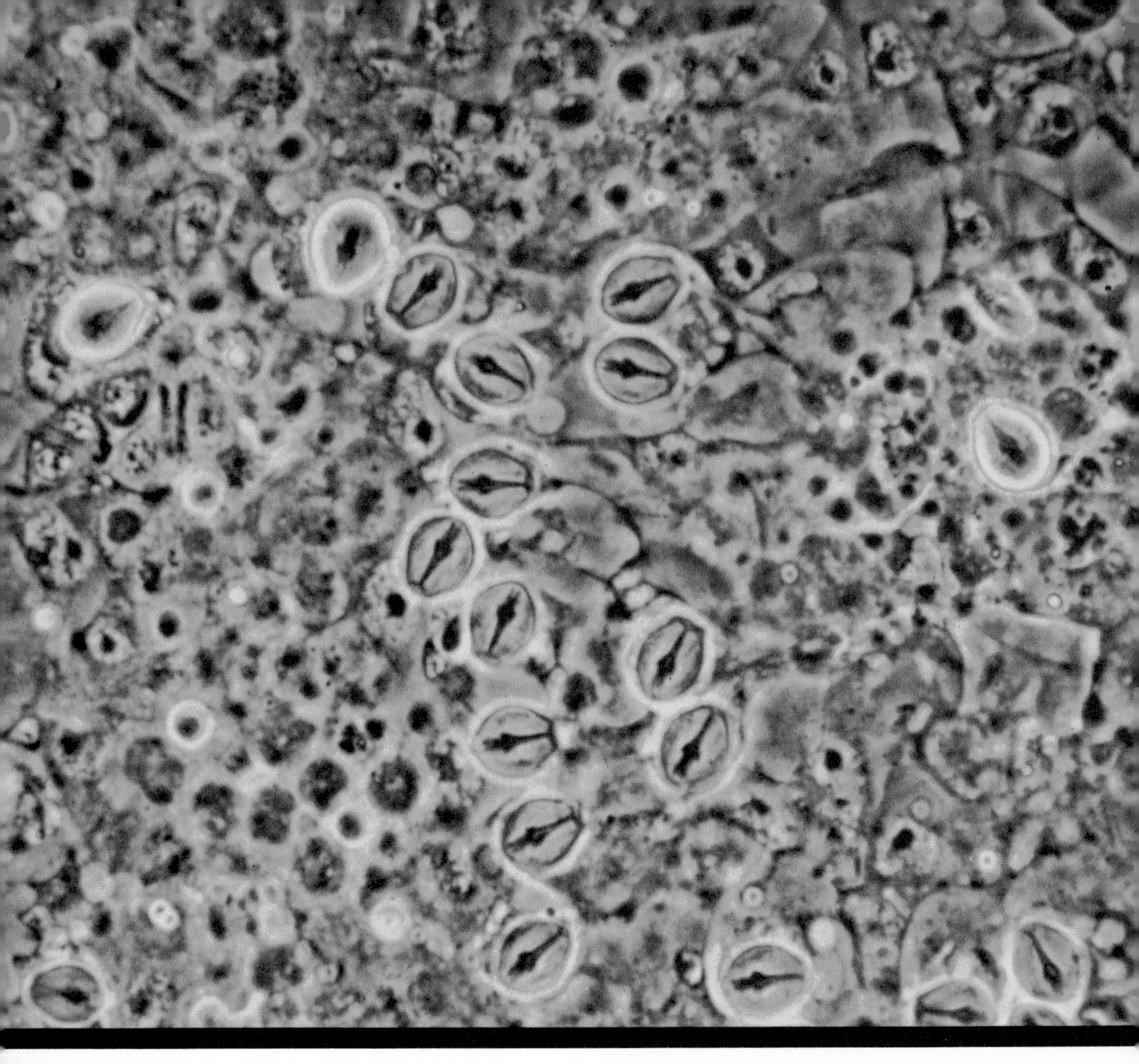

Nematocysts. These globular-shaped cells respond to mechanical stimulation as well as chemical stimulation. When they are triggered, a tiny thread, which can sting, stick to, or wrap around a victim, is forced out.

Stingers

The tentacles of many coelenterates are lined with small stinging cells called nematocysts. When an animal armed with these cells is stimulated by another living organism, the slightest brushing of its trigger causes tiny threads to extrude.

These threads are designed to act in one of three general ways. Some wrap themselves around the triggering object and thus entangle small prey. Others have a sticky coating that adheres to anything it touches. The coating holds the prey until it can be eaten. Another kind of thread has a hollow barbed tip at its end which works like a harpoon.

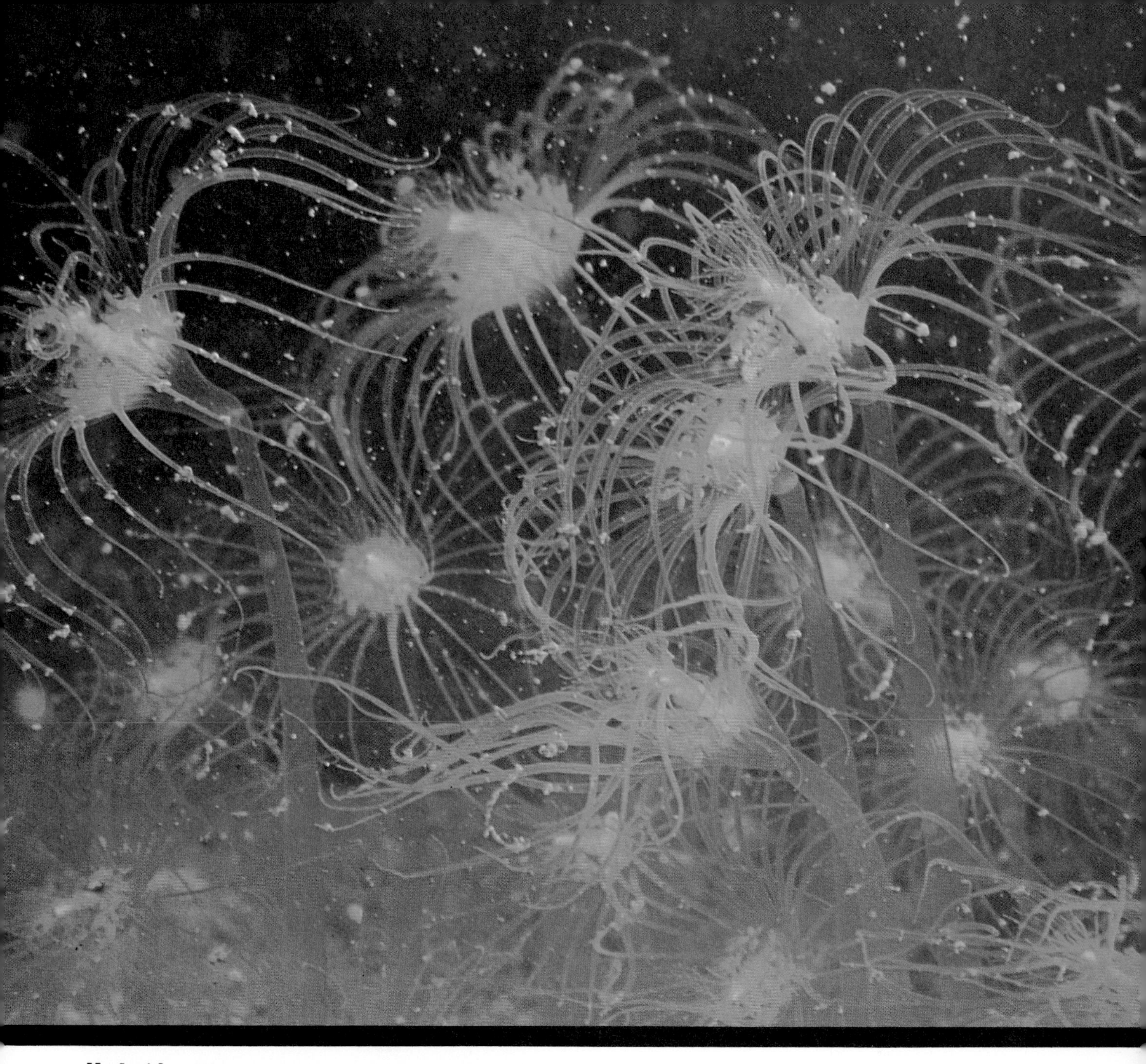

Hydroids. *These animals, looking dainty and feathery, are armed with stinging cells. They are facing the current to capture whatever floating debris or plankton touches their tentacles.*

Generally the small darts can only penetrate the skin of tiny animals. Those of some jellyfish, for example the Portuguese man-of-war, can penetrate the skin of man—with severe and sometimes fatal results.

Another animal that uses poison to obtain a meal is the cone snail. Most snails scrape algae off rocks by means of a tooth-studded, ribbonlike apparatus called a radula. The highly specialized radula of the cone snail is modified to form hollow teeth that are connected to a gland containing a very toxic poison. The snail strikes its victim with the radula, injecting the poison.

Deceivers

Many mechanisms assist both in defense and in food gathering. Camouflage is one of them. Fish camouflaged are able to stalk or wait for their prey.

Flatfish have both eyes on the same side of their head. They reside on the bottom and their coloring makes them difficult to distinguish from their "backgrounds." To add to the disguise, they cover their bodies with a light sprinkling of sand. All that is visible are their protruding eyes—looking for crustaceans, bottom-dwelling worms, and other fish. When something edible approaches, the fish springs to life and consumes the prey.

Flatfish. All the 130 American species of flatfish have both eyes on one side of the head. They are remarkable for their camouflage ability.

If nothing passes by, it swims along for a short distance and settles again on the bottom and simply waits.

Many other fish make use of camouflage. Trumpetfish have elongated bodies which make them difficult to see as they hide among gorgonian corals. They stand on their heads and hang motionless in their hiding places until something nears. The sargassum fish has a body covered with fleshy protuberances colored in such a way that it looks like the seaweed in the Sargasso Sea where it lives.

Borers

The bane of wooden-ship admirals like Lord Nelson was a weird mollusc, the *Teredo navalis* or shipworm, which, if not exactly an inactive animal, certainly may be considered a sedentary feeder.

The tiny larva of this animal swims through the sea until it finds a piece of wood lying under the surface. Here it settles down and develops a small shell—but a shell that is no more than a pair of one-half-inch plates, which it uses as augers even though the mature clam may grow to as long as two feet. With these miniature drills the young shipworm bores into the wood and is able to excavate a long cylindrical hole.

The shipworm can occur in great numbers, so crowded together that only filmlike partitions separate one burrow from the next. When this happens, the infested timber can bear no weight and crumbles. Each hole is

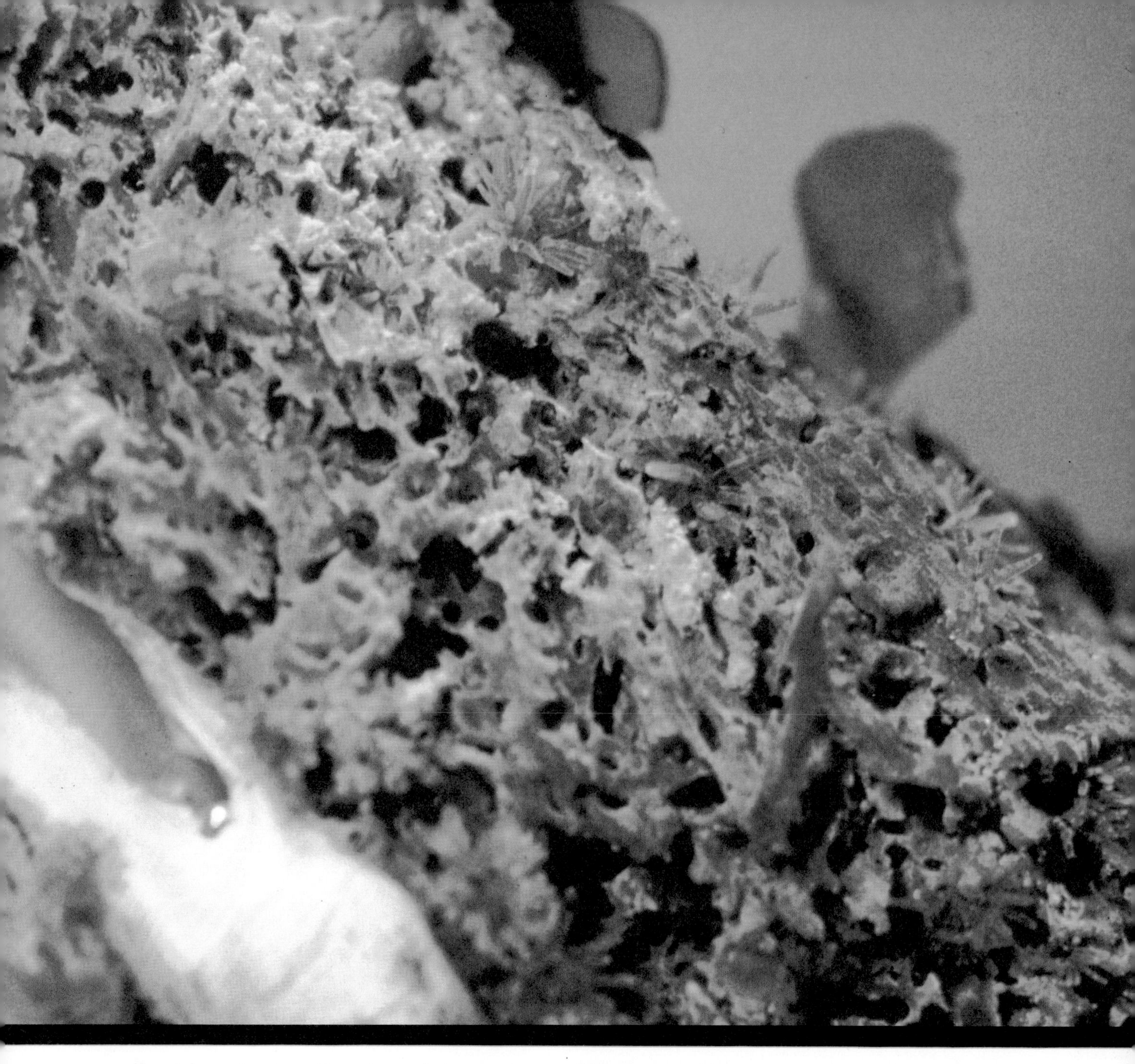

usually driven in the direction of the grain of the wood, but not always. One burrow never breaks into another. What is also curious is that the shipworm never enlarges the original opening to his hole, so in effect he has imprisoned himself in his own burrow—maintaining connection with the outside ocean by means of two miniscule siphons. It is through these siphons that his food comes. He gets no nourishment from the wood he is reducing to dust.

Shipworms. *Shipworms can be considered the termites of the sea. Just as termites can destroy the foundation of a house, so, before the arrival of the iron hull, the shipworm was the nemesis of the dockyard.*

Some notion of the cost of the shipworm's operations can be arrived at by seeing the huge percentage of the total correspondence of eighteenth-century European navies which was devoted to it. Not until the coming of the iron hull was the problem solved.

Triton snail. *Though harmless to man, these large snails feed on very unpalatable prey. Sea urchins and other sharp-spined animals are easy victims of these crawling animals.*

Nudibranch. *Looking like beautiful, exotic flowers, these molluscs seem to live off prey few other animals would touch. The stinging nematocysts of the hydroid fail to bother the nudibranch.*

Slow but Sure

Even the spine-covered sea urchin isn't immune to the law of the sea. Here a triton snail approaches one. Slowly but surely the triton feels out its victim. And just as slowly and surely it envelops the sea urchin. Under the slow and studied onslaught of the triton, the sea urchin is helpless. The nudibranch, also shown here, is another slow feeder. This shell-less snail moves up and over hydroid colonies, leisurely consuming its victims.

Chapter IX. Active Feeders

The relatively tranquil life-styles of the animals we have just considered don't suit all the sea's population. Many animals pursue their food vigorously—swimming long distances, following seasons and population fluctuations, resembling supermarket shoppers in picking up items here or there from the coral or the bottom. To these animals eating is serious business and they work hard at it.

Active feeders vary in size, shape, and in diets. Sharks are long and sleek. Tunas look like torpedoes. Many reef dwellers have deep, flattened bodies and pointed snouts. Bottom feeders have underslung mouths. Some deep-sea fishes have mouths capable of holding fish larger than themselves. Talented swimmers like the tunas, jacks, and blues rely on great bursts of speed. They generally attack a school of smaller fish, scattering its members. Their streamlined, powerful bodies can outswim but not always outmaneuver smaller creatures.

"Many animals pursue their food vigorously—swimming long distances, following seasons and population fluctuations, resembling supermarket shoppers in picking up items here or there from the coral or the bottom."

Many animals feed whenever an opportunity presents itself. But the habits and feeding cycles of most predators are timed to the habits of their prey. Animals active during the day are preyed upon by diurnal predators. The same is true for nocturnal creatures. This tends to spread the action and ease competition for food. The millions of tons of light-shy animals that comprise the deep scattering layers come to the surface at night to feed on plankton.

The feeding habits of marine animals are reflected in their food-capturing mechanisms. Spikelike teeth can grab and hold a wriggling prey. The broad flat molars of manatees are suited to grind up plants. Sharks that take big chunks of large animals are equipped with sharp triangular teeth with serrated edges. The points sink into tough skins and the sides cut the flesh. Fish that scrape algae off rocks and coral have flat teeth with cutting edges similar to our own front teeth. Some animals don't have any teeth at all; they include the largest mammals on earth, the whales.

In a school opportunities for catching a good meal would seem promising. But it is not as easy as it looks. When a jack attacks, the wall of bodies suddenly breaks up and small fish disperse in all directions in a confusion of motion. A jack must select a single fish from the wall and concentrate all his attention on it. Most predators catch their prey tailfirst; then with subsequent bites and rejections the prey is manipulated into a headfirst position, where all the fin and body spines point harmlessly backward. It can then be swallowed.

The appetites of these active feeders do not drive them to exterminate the species on which they prey. There exists a balance between the number of prey and the population of predators they sustain. Often, the weakest or least fit of the prey are the first to be captured.

Jacks. *Although there are many varieties of these hungry fish, they all have one very important thing in common. They are very fast—both in the way they travel and in the way they capture their prey.*

Reef Feeders

Many fish which live on and around coral reefs differ from their open-ocean counterparts. Their bodies appear too deep for their length, and quite flattened, giving them a disclike appearance. However, this design is suited for maneuverability in the restricted environment they inhabit. Fish that feed on small plants and animals found in narrow cracks and small holes in the coral have elongated snouts. The teeth of the reef dwellers are specialized, too, varying according to the fish's diet. Some use teeth to pluck animals from the reef, while others have piercing teeth which grasp and hold prey. Surgeonfish are herbivorous reef inhabitants. They usually feed on algae growing on rocks and coral. In most species the teeth have become sharp and flat-faced, which makes them well-suited for scraping up tiny plant life.

Feeding by Suction

On one of our voyages we moored *Calypso* over an unspoiled reef north of Madagascar. Here we had the good fortune to make friends with a large grouper we named Ulysses. When we were about to depart we wondered whether Ulysses might have become dependent on our handouts. We needn't have worried. We knew Ulysses was as agile as ever when we saw him in action on a big fish we'd speared for *our* dinner. Left to their own devices, groupers wait in lairs for their meals. When a mullet, grunt, or crustacean comes within range, the grouper snaps his mouth open. The vacuum created by the open mouth literally sucks in the victim. If the prey is too large to fit the mouth, its tail will stick out and be held between small teeth and fleshy lips until the grouper's pharyngeal teeth have dealt with the victim's head.

Feeding in the Abyss

Half of the earth's surface is covered by more than 13,000 feet of water. It is the abyss—vast plains interrupted by volcanoes, rugged mountain ranges, and great scarring fractures. No man has walked here as he has on the moon. Only a few men in bathyscaphes have visited it. Our knowledge and understanding of life in this hostile world is derived almost entirely from a limited number of automatic photographs or samples taken almost at random in trawl nets.

Most of the abyssal inhabitants are believed to take their nourishment from or on the mud or by filtering it from the water. Sponges pump water. The bottom is covered with plow marks of burrowing sea cucumbers and worms. Many deep-sea fishes have large heads, tapering tails, and long anal fins. The body form gives the fish a downward thrust as it swims forward and the underslung mouth sifts through the bottom ooze. Even the mighty sperm whale sometimes feeds in this manner.

Many unusual anatomical systems have developed in deep-sea fish for food gathering in this impoverished zone. To many deep-sea fish, lights are important in the eternal darkness. Deep-sea anglerfish have luminescent organs located in or above their mouths to lure unwary prey. Many deep dwellers are known to possess sharp eyesight. Barbels under the mouths help fish to locate a meal in the mud. Some abyssal creatures, like the squid, rise toward the surface at night to feed in the rich upper waters.

Opportunities for predators occur infrequently. One family of deep-sea fish is equipped to handle almost anything that comes along. They are called the stomiatoids, scavengers of the deep, capable of engulfing dead creatures several times bigger than themselves. Most of these fish have barbels; one of them trails a chin whisker ten times its own length. The barbel in this fish is not used to detect food in mud, for these are not bottom feeders. William Beebe observed that motion of water near the barbel caused one of these fish to thresh about snapping its mouth. So it would seem to be a sensory organ.

"The abyss—vast plains interrupted by volcanoes, rugged mountain ranges, and great scarring fractures. No man has walked here as he has on the moon."

A typical member of this group is the dragonfish, an animal that grows to a length of 12 inches. Unlike the usual carnivorous fishes that eat prey smaller than themselves, or those that take bites from bigger animals, the dragonfish is able to take whole prey of its own size or larger. It does this with a spectacular arrangement of teeth and jaws and a stomach capable of enormous expansion. In action, the upper jaw is extended vertical to the body axis and the lower jaw juts up and forward so that both upper and lower teeth penetrate the victim and hold it fast. This elaborate system provides the dragonfish with some important advantages. The large mouth increases its potential for capturing both larger and smaller fish. With a large mouth and a small body, the dragonfish requires relatively less food.

Deep feeders. *Stomiatids, the family of fishes which includes the dragonfish, all have large eyes and expansible mouths, with large sharp teeth in both upper and lower jaws. They also have luminescent organs which are used for attracting prey and finding mates.*

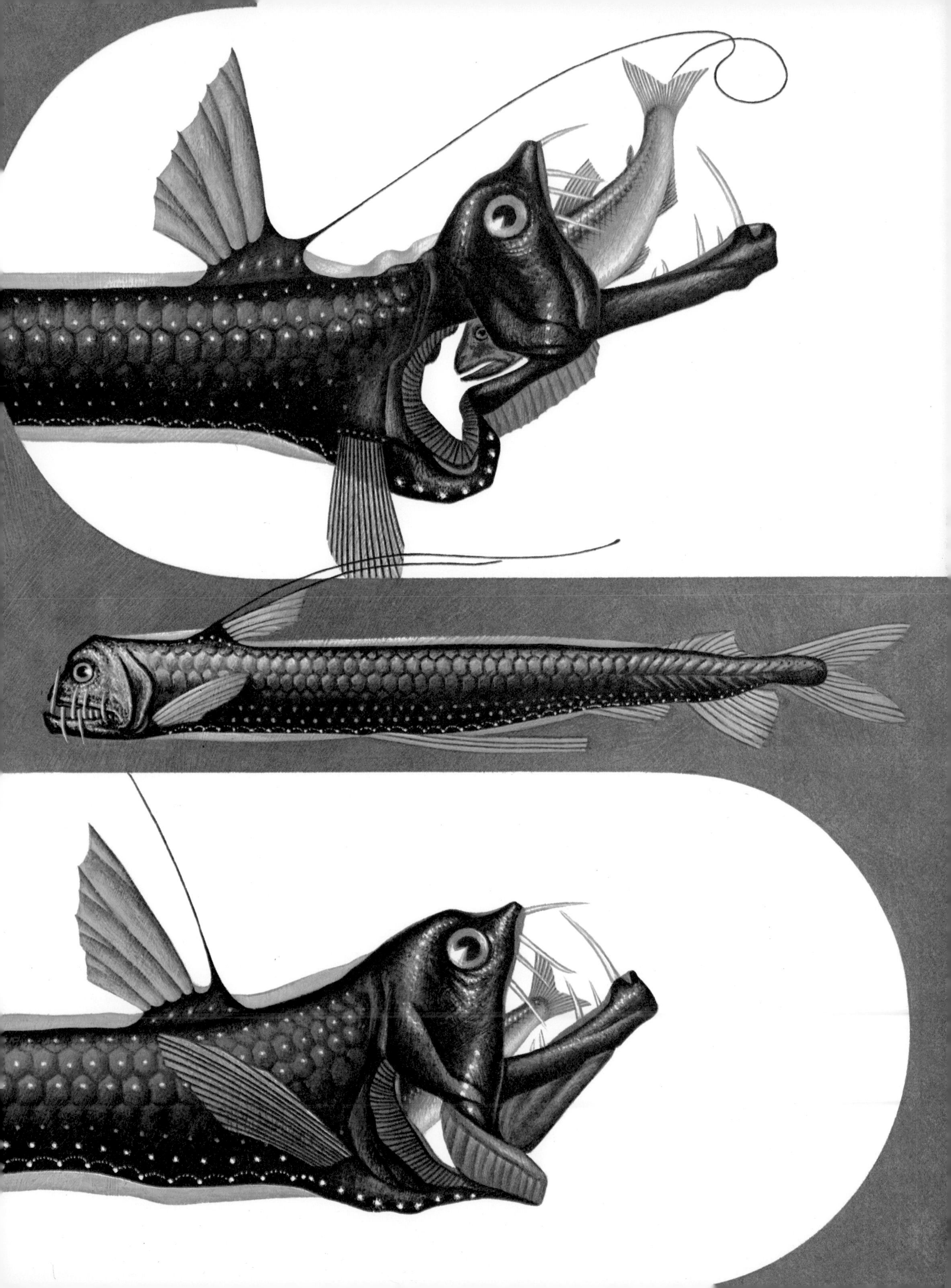

Stunning Their Prey

The torpedo ray is a shocking animal. It has electric organs near its head in each wing capable of producing a charge of up to 200 volts, enough to knock down a man. When aroused, the animal sets off a series of electrical discharges of descending power. Once spent, the battery takes several days to recharge. Having electric organs helps the torpedo ray find and capture prey. Some members of this family are blind, all are sluggish swimmers. Other rays feed on fixed or fairly immobile animals. But the electric ray is known to eat active fish. It is possible that the ray emits weak voltage pulses that act as a form of detector, advising the animal that prey are near. When a codfish, for instance, wanders into its electrical field, the ray jumps from the bottom, enfolds the fish in his wings, and stuns it. Then he can eat at leisure.

Upheaval and Vacuuming

The gills of the bat ray are located on the underside of the flattened body. As the fish swims over the sea floor, a stream of water spurts from its gills, stirring up sediments and uncovering the animals hiding there. When the ray discovers an abundant food area, it stops its forward motion and settles gently to the bottom. The greatly enlarged pectoral fins begin to flap, whipping up a current. A cloud of loose sand and mud lifts from the bottom and in the cleared area large numbers of bottom dwellers are exposed. The ray eats them. Occasionally a shellfish, firmly attached to the bottom, refuses to yield to the hungry ray. To pry the animal loose, the bat ray presses its supple body over the creature, forming a vacuum between itself and the bottom. The thrust of the vacuum breaks the hold of the prey which then is sucked into the ray's mouth.

Plant Eaters

The marine iguana of the Galápagos Islands, dugongs, and manatees, all air breathers, are descended from animals which once were land dwellers.

The marine iguana looks like its land-dwelling relative, but his feeding habits differ radically from the terrestrial form. On the barren Galápagos Islands food is scarce, but the iguanas are able to graze on algae growing below water. Today the marine iguana can roam beneath the sea for as long as one hour, munching on these small plants with his sharp tricuspid teeth like a dog gnawing on a bone. He is able to control his heart rate, slowing it from 45 beats per minute on land to as few as four to five in the sea. By doing this he conserves oxygen. He can even stop his heartbeat completely for as long as three minutes. The ability to regulate the use of oxygen increases the time he can spend below, and thus he can browse selectively. When he is submerged, his bloodstream borrows oxygen from body tissue.

Being cold-blooded, these diving lizards have no choice but to take on the temperature of the frigid waters surrounding the Galápagos Islands. They emerge with a body temperature below what is comfortable for them, and as a consequence, they must warm up. This is accomplished by their temperature-regulating behavior. When cold, they lie on the rocks at an angle which puts them perpendicular to the rays of the sun. As their body temperature rises, they reposition themselves so as to absorb less of the sun's direct rays. If they are too warm, they sit facing the sun, parallel to its rays. In this way they are able to achieve some degree of temperature regulation.

The manatees and dugongs are farther along in the process of returning to the sea. These mammals probably descended from a four-footed animal which roamed primordial marshes browsing on the lush vegetation. For unknown reasons certain groups of this herbivorous species ventured farther and farther into the water in their search for food, and passing time has brought about many changes in their descendants. They have lost their hindlimbs and their forelimbs have changed into flippers (retaining vestigal toenails). Their body form has been altered and their reproductive habits adapted to an exclusively marine environment.

> **"The marine iguana is able to control his heart rate, slowing it from 45 beats per minute on land to four to five in the sea. He can even stop his heartbeat completely for as long as three minutes."**

The manatee's tremendous appetite has been of great help in Florida in controlling the spread of water hyacinths. This lovely but troublesome plant was introduced to these waters from Africa, and because it knew no predators it thrived and today chokes many waterways. The manatee has acquired a taste for these plants and by eating them he helps us in two ways: his services cost us nothing and are ecologically less harmful than the herbicides used to control these weeds.

Iguana. *The marine iguana of the Galápagos Islands (top) is a good swimmer and has strong claws for grasping the submerged rocks against the surge while it is underwater grazing on algae.*

Manatee. *This large, docile vegetarian (bottom) lives a slow-paced existence, rarely venturing out to the open sea. He remains in and around rivers and their estuaries, where vegetation is easier to find.*

Gill rakers. In the drawing above, the red parts of the gills are gill filaments through which oxygen is transferred into the bloodstream of the fish. The light-colored parts of the gills are the gill rakers, which are used in some fishes to filter minute plankton passing through. In others they serve as protection for the more delicate filaments.

Herring. At right are a group of herring. These fish are filter feeders and use their gill rakers to filter plankton from the water.

Gill Raker Mechanism

In some fish such as the tuna, shark, and bluefish, the gill rakers are short, stubby, widely spaced projections on the gill arch. As water and other materials enter the mouth, the rakers serve as a protective device, preventing large chunks of food or debris from passing through the gill openings, reaching the gills, damaging the delicate filaments and affecting the fish's breathing capacity. Other fish like herrings and manta rays have more elaborate gill raker mechanisms. In these fish the rakers are long, slender, close together, forming a fine mesh which strains minute food particles or small fish from the water.

The individual rakers vary in design. Some are simple smooth shafts and others more complex with fine feathery appendages. As the design becomes more complex and the number of appendages increases the strainer becomes more efficient and is able to sift ever finer particles from the water. The system is remarkably efficient: the 12-inch menhaden strains approximately 6.8 gallons of water per minute and from this volume several cubic centimeters of plankton concentrate are filtered out, primarily diatoms and small crustaceans. The gill raker system for collecting food is fairly common.

Filter Feeder

In a series of languid loops—a three-dimensional ballet—the manta ray feeds. This large graceful animal, cousin of the sharks, is a filter feeder that passes many gallons of water through its huge mouth each minute. Two feeding fins on either side of the mouth are directed forward and fan fish and plankton into it. In the throat a screen of small, spiny protuberances holds the food until it can be swallowed and keeps it from clogging up the animal's gills. In action the manta's feeding fins look like horns and have caused is to be given the name "devilfish" and a reputation it doesn't deserve. When we began diving, fishermen warned us that mantas killed divers by wrapping their wings around them and smothering them. This is not true. But their great size (one with a 22-foot wingspan has been measured) attests to the efficiency of their feeding methods.

Large Fish—Small Prey

The whale shark, largest fish on earth, swims lazily about at great ocean depths. In my entire career I have encountered one only five times. As the whale shark swims about, its mouth is always open, like the intake of a jet engine. The food is filtered from the water through an elaborate gill raker system. The mouth stretches to a width of six feet and opens about 18 inches. Unlike other sharks whose mouths are under the head, the whale shark's mouth is located at the tip of his body. In spite of its diet of minute foods and the efficiency of its gill rakers, it is equipped with approximately 310 rows of small teeth in each jaw. These are probably used to crush prey too large for easy consumption. Whale sharks are quite docile. We have ridden on their backs without fear, although their teeth are formidable and a stroke of the tail could spell disaster.

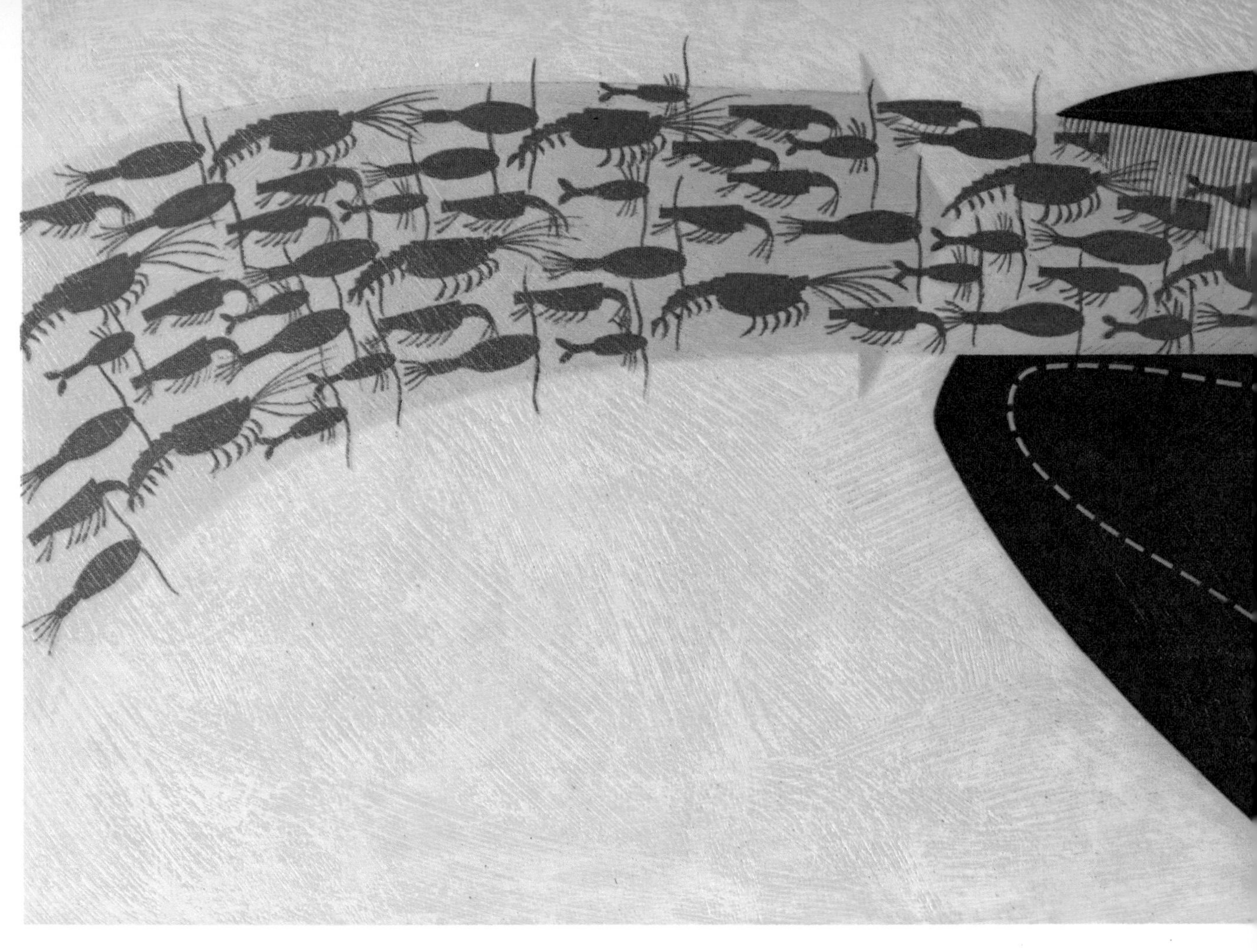

How the Baleen Whale Filters Its Food

Many whales, including the biggest, have no teeth with which to chew their food. They are equipped with another feeding mechanism suited for ingestion of huge quantities of tiny fish and crustaceans, especially krill.

When filter-feeding fish extract plankton, they allow water to enter their mouths and exit through their gills. The food is strained out by gill rakers. Whales lack gills and gill rakers. Water entering their mouths must also exit from them. So the teeth of the whalebone whales have been replaced by efficient strainers called the baleen. (The fetuses of baleen whales have tooth buds, but these rudimentary teeth disappear when the baleen develops, and no trace of their presence is found in the adults.)

The baleen apparatus varies between species, but the general principle is common to all. There may be as few as 250 plates or as many as 400 hanging from each side of the upper jaw. (The baleen of "whalebone whales" is not made of bone but of keratin, the same material our hair and fingernails are made of.) The plates are roughly triangular in shape, wider at the top than at the tip, and about one-fifth of an inch thick, separated by a space of one-half inch or less. The edge in contact with the inner surface of the lips is smooth; the edge facing the inside of the mouth is frayed into strands of coarse hairlike material. These hairs become entangled with one another, forming a thick fibrous mat. This mat is the real strainer. The length of the baleen plates varies from about eight inches to about 14 feet.

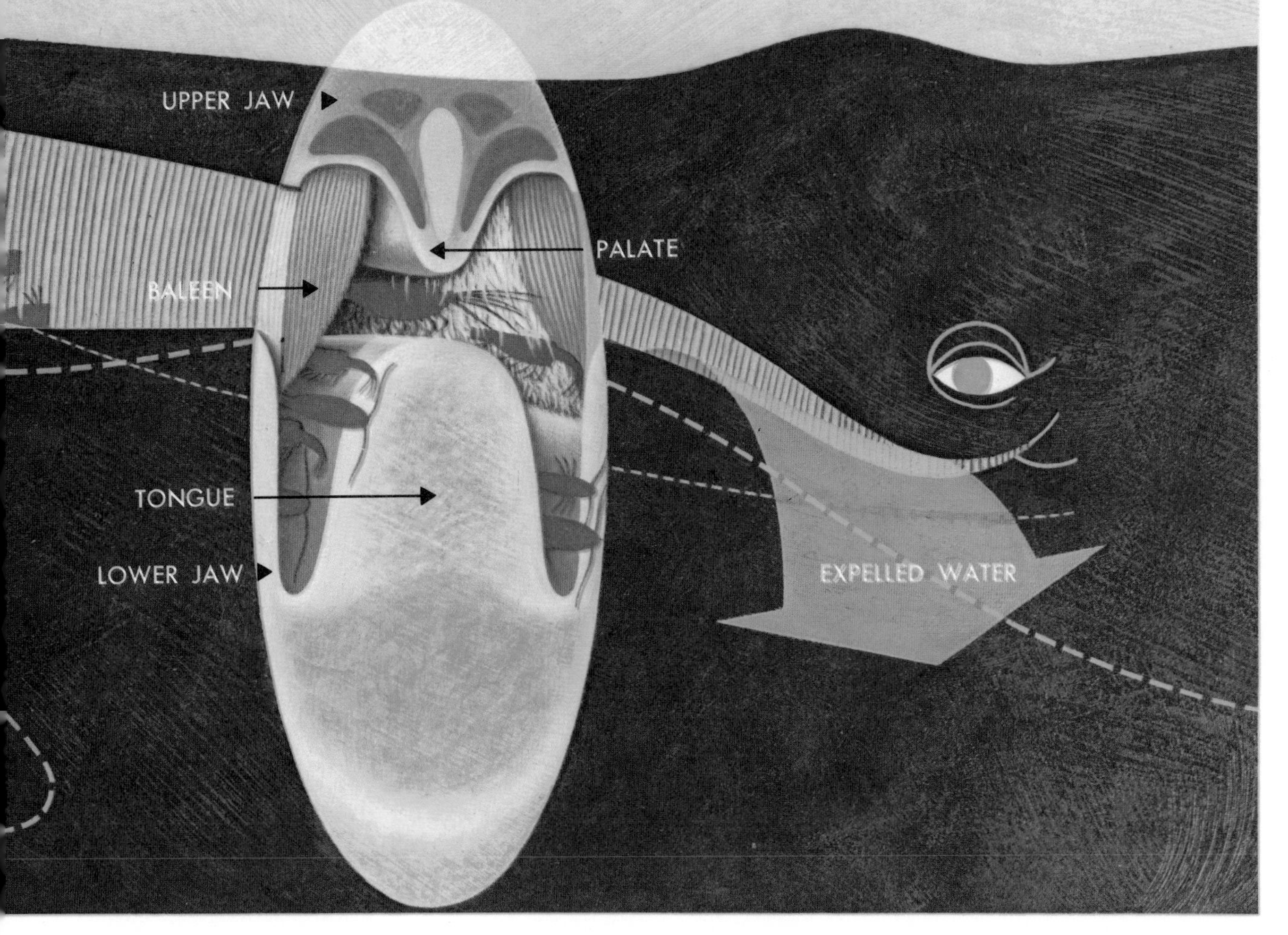

No diver has ever witnessed the actual mechanics of feeding of any baleen whale. But today we understand that these mechanics, as well as the nature of food obtained from the sea, vary from one species to another according to the anatomy of the mouth. Generally speaking, whales seek rich layers of the ocean within their diving capacity.

The two extremes are the slow, chunky right whale and the fast, sleek finback or blue whale. The right whales are characterized by the exaggerated arch of their upper jaw or rostrum. When it is fully opened the long baleen still reaches to the base of the mouth; when it is closed, a great lip five feet high reaches up from the lower jaw to seal the mouth. They probably cruise at four or five knots, with their mouths open nearly continuously. Plankton-rich water flows right through their mouths, exiting near the rear corners, the krill trapped inside.

The baleen. *Zooplankton are taken by the feeding whale and strained out by the bristlelike baleen —and then swallowed. The vast quantity of water taken in with the zooplankton is expelled.*

On the other hand, finback and blue whales have longitudinal folds along the lower jaw and a throat that can be blown out when the mouth opens to huge proportions, taking in up to 15 tons of seawater. As these whales cruise at 12 to 15 knots, any sizable fish or squid can be trapped as well as krill.

"Right whales probably cruise at four or five knots with their mouths open almost continuously. Plankton-rich water flows right through their mouths, exiting near the rear corners, the krill trapped inside."

Utensils and Sensors

Sea otters require huge amounts of food daily—considering their size. They weigh from 45 to 100 pounds but generally eat eight to 15 pounds of food a day. To satisfy their needs, about half their waking hours are spent searching for food (fish, molluscs, sea urchins). The search is facilitated by the sea otter's ability to stay submerged for up to four minutes and dive to depths as great as 200 feet.

Lacking sharp cutting teeth to open stubborn shellfish, the sea otter would appear to be unsuited to a hard-shelled diet. However, otters are almost unique in having learned how to use rocks, much as we use tools, to assist them in their efforts to get at the succulent meat of hard-shelled animals. When an otter dives to the bottom to collect an armful of mussels, crabs, snails, and sea urchins, he also picks up a large rock. Once safely back on the surface he rolls onto his back. Looking like an old man reclining on a beach float, he begins to dine. The rock is settled on his chest and used much as an anvil is. He places a shellfish in his hand and smashes it against the rock again and again until it breaks open. Then he pries the shells apart and picks out the meat, tossing the empties aside. Every 30 seconds or so he clutches the rock and uneaten food close to himself and rolls over, fastidiously washing any food debris from his gleaming coat.

"Looking like an old man on a beach float, he begins to dine. The rock is settled on his chest and used much as an anvil is. He places a shellfish in his hand and smashes it again and again against the rock until it breaks open."

Sometimes the shellfish are firmly attached to the bottom, and the otter can't pick up his meal. Then the otter uses his rock as a hammer and pounds the shellfish until its viselike grip on the substrate breaks. The *Calypso* team recently made a breakthrough in understanding these remarkable creatures at close hand. A bond of trust was painstakingly developed between the gentle, careful aquanauts and the wary otters. The patient divers were eventually able to feed the otters by hand.

Walruses dive into cold arctic seas and descend to depths of as great as 300 feet in search of food. It is still unknown if their tusks actually play a major role in food gathering; in large males these may reach lengths of from 14 to 20 inches. Like a great two-tined rake, a walrus's pair of tusks could drag through the bottom, turning over mud and rocks, exposing clams and worms. But we have no evidence that this behavior takes place. The lips and whiskers are capable of digging shellfish out of the mud; broad flat molars crush shells. The animals isolate the liberated meat and swallow it, rejecting shells and stones. Inevitably some stones and shell fragments are swallowed; they are later regurgitated. But what we know is that the great tusks are used as efficient defensive weapons against such predators as polar bears or killer whales. They are also used in intimidation fights in the mating season.

Sea otter. *In addition to eating crabs and abalones, these mammals (top) eat sea urchins, which eat kelp. Sea otters around the kelp beds of southern California, therefore, have insured the survival of the kelp. Kelp's greatest enemy today is pollution.*

Walrus. *These huge animals (bottom), which can grow to a weight of 3000 pounds, must eat large quantities of food each day. Their diet includes clams as well as vast quantities of ascidians, crabs, and even starfish.*

Big, Fast, and Hungry

The slashing teeth and speed of this eight-foot Australian shark are too much for the 15-pound porgy that it grasps in its mouth. Sharks know few predators, except other sharks, but they fear orcas, whales, and dolphins. The muscular, streamlined bodies, sharp teeth, and powerful jaws of the more voracious species permit them to make the most of a battery of finely tuned senses. A highly developed ability to detect odors and vibrations makes them especially effective in tracking down animals that are injured or otherwise distressed. Thus a wounded fish whose body fluids have been released into the water, or that swims erratically, is a prime target for sharks. In contrast to the aggressiveness of these more active species, many other sharks are sluggish yet efficient creatures that subsist on shellfish and other sedentary prey.

Intelligent Hunters

The orca, a close relative of the dolphin, also uses a highly sensitive sonar system to locate and identify prey. These toothed whales eat fish and mainly squid, but occasionally enrich their diet with meat of warm-blooded animals. Often hunting in packs, they attack much larger whales when these are sick or wounded, and using their teeth reportedly snap off the pectoral fins and tongues of their toothless and more-or-less defenseless relations. They also catch seals, dolphins, and even penguins, which they swallow whole. A single stomach of a dead orca contained the remains of 13 dolphins and 14 seals; traces of 32 adult seals were found in another. Orcas are intelligent and are known to bump the undersides of ice floes in an attempt to tumble into the sea the animals riding them. Their jaws resemble those of a huge trap with about 50 teeth.

Chapter X. Feeding Relationships

Among the most popular and useful animals in the sea are the little creatures that specialize in grooming others. The "others" are larger animals who present themselves to be cleaned of parasites and diseased flesh. Relationships of this kind are known as symbiotic. There are three main categories of symbiosis: mutualism, commensalism, and parasitism. In mutualism, both of the participants benefit. Commensalism describes an association in which one partner gains an advantage and does not harm the host. A parasite lives at the expense of its host, in extreme cases killing it.

It appears that the cleaning service is a vital necessity for marine animals. When those capable of carrying out the chore were removed from an isolated reef all but the most territorial fish left the area and those remaining were in poor condition.

In another kind of mutualism, that of the hermit crab and the sea anemone, neither can survive alone. The hermit crab with its small shell benefits from the anemone, which provides a protective covering. The anemone gets its food and transportation from the crab. Remoras with special suction discs developed from the dorsal fin, allowing them to attach themselves to large animals like sharks, mantas, and turtles, share their host's meals whenever it is feeding. They repay their hosts by ridding them of parasites.

"The clownfish, less than six inches long, is believed to secrete a special mucus which prevents the anemone from discharging its lethal stinging cells. Sea anemone and clownfish are truly partners—living in what scientists call symbiosis."

Symbiosis can either enhance or diminish the reproductive capabilities of marine animals. The female squid attracts a mate by flashing borrowed luminescence. She has a special nidamental gland in which she grows and nurtures the luminescent bacteria that help her get her man. On the other hand, the root-headed barnacle, *Sacculina,* destroys its host's reproductive ability. This animal invades crabs to take nourishment from their blood and incidentally alters their sex hormones so that males at their next molt assume female form.

Another example: a few small creatures enjoy complete immunity from the tropical sea anemone. The clownfish, less than six inches long, is believed to secrete a special mucus which prevents the anemone from discharging its lethal stinging cells. Sea anemone and clownfish are truly partners—living in what scientists call mutualism. So close is their association that aquarium studies have shown the anemone unable to live without its associate. Why we are not certain. The anemone's feared tentacles provide the clownfish with a hiding place from predators; it also feeds upon food scraps left over by the anemone. Vividly colored, the clownfish may lure other fishes within the anemone's grasp. When hungry, it wanders out to food sources, feeding until its hunger is sated. Then, if it has some food left over, it fulfills its part of the mutual arrangement by bringing some back to the anemone.

Clownfish and anemone. *These animals live in harmony. The stinging nematocysts of the anemone protect the clownfish from predators, while the clownfish may bring food to the anemone.*

The Algae-Coral Relationship

The delicate green tracery seen on reef-building corals is a sign of the continuing relationship between the coral polyps and green plants which might have begun some 450 million years ago. Today corals and green algae form the basis of the most complex marine ecosystem. Corals are animals that feed on plankton and extract calcium from the water to build their rocklike skeletons. They do not engage in photosynthesis; still, they can exist only in sunlit waters. They are efficient at building their skeletons because of a symbiotic arrangement they have with unicellular green algae living within them. There is an exchange between these two organisms: the plant photosynthesizes compounds which nourish the coral and in the process gives off oxygen; in turn, the animal's waste products furnish elements needed by the plant for growth.

Calling the Cleaners

Fish queuing up at the cleaning station indicate their desire for the cleaner's services in a number of ways. Some present their heads to the cleaner. If a particular problem troubles them they may direct him to the affected part of their body first. Sometimes the customers assume unusual postures, presumably conveying messages to the cleaner. Blacksmith fish vying for the attention of a female may position themselves vertically, with heads or tails pointing upward. Some blacksmiths lie on their sides, others turn themselves upside-down. The goatfish, shown above, signals its cleaner, the butterflyfish, by flaring its pectoral fins and extending the barbels hanging from its chin. It also can change from a pale color to a reddish pink, and the goatfish uses this device to signal the cleaner and aid him in locating parasites and diseased tissue.

Wrasse cleaning bass. *When time for a cleaning comes around, the bass will go to a cleaning station and await the arrival of a little wrasse. The wrasse, while the bass remains still, will clean the fins, gills, and even the mouth of the larger fish.*

Shrimp cleaning moray eel. *Like the bass with its wrasse, the moray eel remains perfectly still while the shrimp picks minute parasites from it. The eel gets cleaned, the shrimp gets fed, and they live harmoniously while services are being rendered.*

Cleaning Stations

Today we recognize more than two dozen species of cleanerfish, six species of cleaner shrimp, one species of cleaner crab. We are sure to find more animal cleaners. They turn up in both tropical and temperate waters, but those in the colder waters tend to be drab in color and fewer in number. They have not specialized to the degree of their tropics-dwelling counterparts. When no customers are around, tropical cleaners signal their occupation with elaborate dances, or they may rush back and forth across the area of their "stations" inviting customers to appear. They are characteristically bold. When they detect one, they take the proper position at the station so the customer won't be kept waiting. The cleaners take great liberties with their customers' bodies, enthusiastically performing their duties by reaching deep into mouths and even probing among delicate gills.

Nearly all societies have imposters; they are found here, too. Some fish closely resemble cleaners both in color and shape; these imposters may even mimic dances to attract customers. But once a mark has arrived the masquerade ends. The fake cleaners do not eat the parasites and diseased tissue. Instead, they nip pieces of living tissue from the unsuspecting customers.

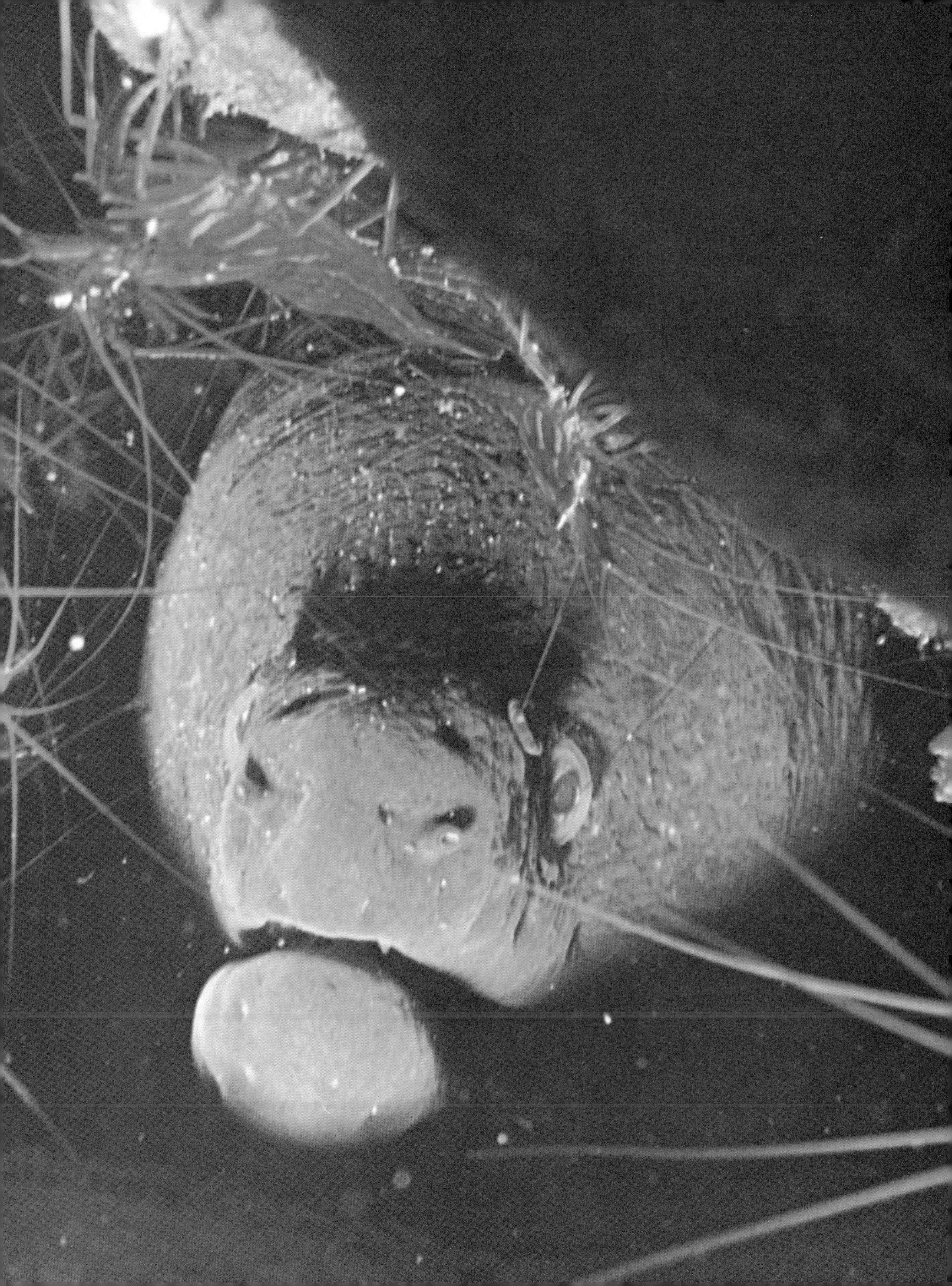

Deadly Protection

The Portuguese man-of-war (a siphonophore) is not one animal but a colony of 1000 or more individuals banded together for mutual survival. Each is specialized, contributing its service to the group: locomotion, flotation, reproduction, food gathering. Cells on the tentacles equipped with tiny, powerful stinging darts (nematocysts) are built to discharge venom. One small fish, the blue-and-black-striped *Nomeus,* spends its life among the dangerous tentacles of the Portuguese man-of-war. It would seem that the *Nomeus* has developed an immunity to the powerful stinging cells, or that the man-of-war is content with their mutual relationship. It is possible that the *Nomeus* might serve as a lure for a predator to attack, whereupon it would be captured by the tentacles of the Portuguese man-of-war. This has not been proven, though.

A Permanent Bond

In their larval stages barnacles drift about the oceans as minute members of the plankton community. After several weeks of this the larvae settle and begin to adopt the sedentary life-style of their parents. Most drop to the bottom where they come to rest on a rock, piling, or shell of a crab or snail. Some species of barnacles attach themselves only on the large bodies of whales; if they fail to find a whale—they die. Once on a whale, the barnacle secretes a cementlike substance, which forms a permanent bond between the huge mammal and itself. When the barnacle finally matures into the adult form, complete with a castle of six calcium carbonate plates, it consumes a wide variety of food that it is exposed to as the whale courses over the wide expanse of the oceans. Older whales, of course, carry more barnacles than younger, smaller whales.

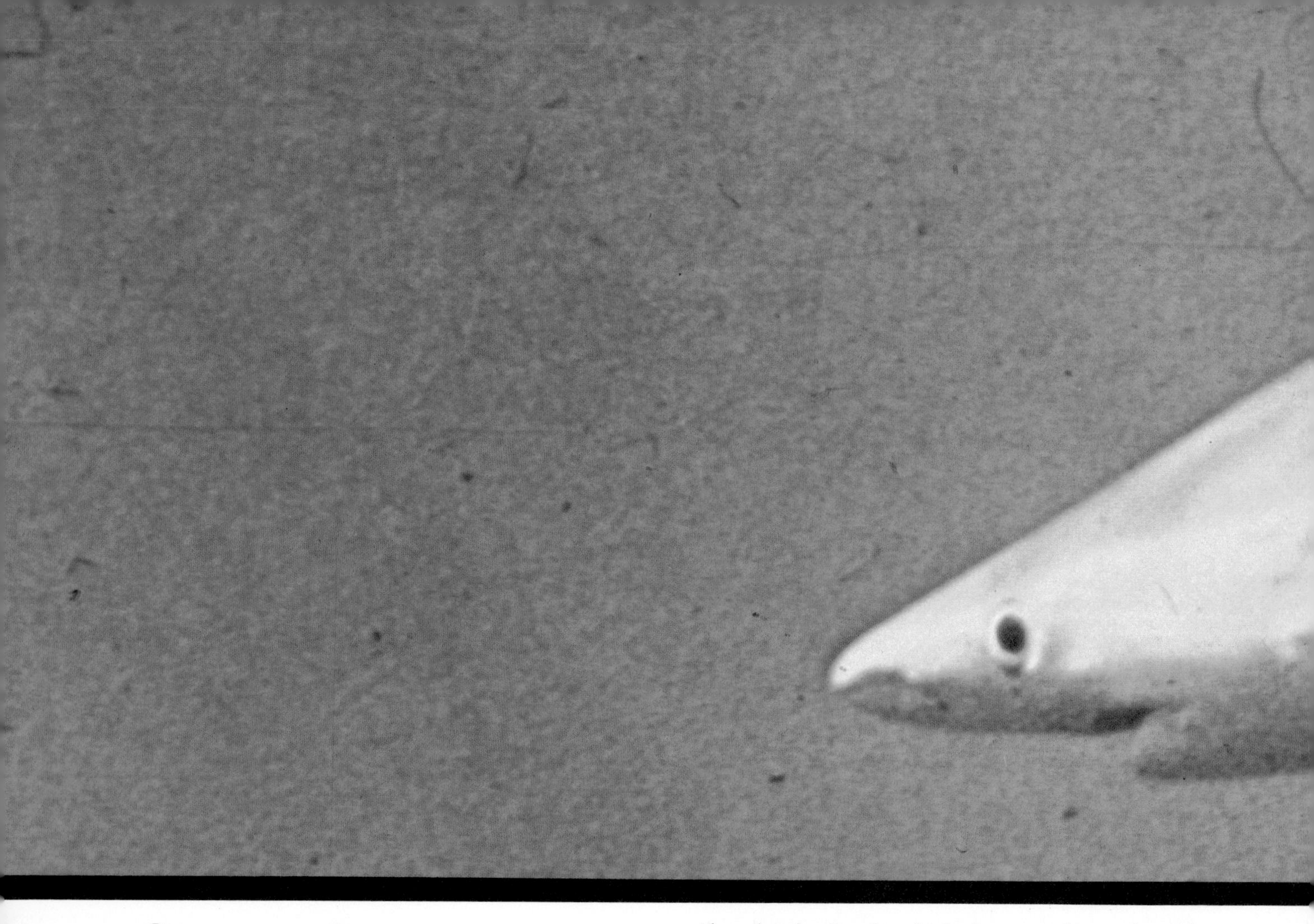

Commensalism

Small, vertically striped pilotfish enjoy close associations with sharks and other giant-sized animals, often swimming in large numbers around them, frequently directly in front of their snouts.

Sharks have poor eyesight, and legend has long held that the little fish guide their companions to prey. The story is unfounded. But once I and one of my earliest diving companions had second thoughts. Frederic Dumas and I were diving in deep, turquoise waters of the Atlantic off the coast of Africa. From the depths suddenly appeared a stout gray shark, unlike any we had seen before, escorted by three pilotfish which followed the shark's movement so closely they seemed to be a part of him. One, no longer than an inch, kept station over the shark's nose. As the shark slowly circled us again and again we felt no concern; other sharks we had encountered had always lost interest in us after a few minutes. I was photographing Dumas swimming in the wake of the animals when the circles began to get smaller and the shark closed in on me. Dumas returned to my side and the tiny pilotfish darted directly toward him. It did not have far to swim, but the time it took him to reach us was sufficient for us to recall the legend. The little fish flitted in front of Dumas's mask like an insect at-

> "Sharks are messy eaters and many scraps of meat drift free when they tear great chunks of flesh from their prey. Littler fish are quick to scurry around for this free meal."

tracted to light. Eventually the shark did make a pass at us but he was fended off by a blow from the heavy camera. He returned to the depths with his entourage of pilotfish when a launch from our ship came to retrieve us.

__Little companions.__ The little striped pilotfish swim alongside, under, and in front of the shark, waiting for a morsel of food to fall their way. This type of symbiotic relationship, in which the shark does not profit but the pilotfish does, is called commensalism.

Actually, pilotfish station themselves in front of sharks and mantas to take advantage of pressure waves formed as the hosts move their large masses through the water. On them the little fish hitch a free ride by surfing on underwater waves, as humans and seals do on surface waves. Traveling this way they conserve energy, not having to exert themselves to keep pace with their powerful companions. Pilotfish also conserve energy by not having to seek and kill for their meals. Sharks are messy eaters and many scraps of meat drift free when they tear great chunks of flesh from their prey. The little fish are always quick to scurry around for this free meal. As far as we know sharks receive no benefits from the pilotfish.

Streamlined remoras are a bit more forward in their associations. They actually attach themselves to larger animals with a large oval suction disc, a modified dorsal fin. Having control of the suction, they can release themselves whenever they desire. When a shark eats the remoras drop off and gobble up food scraps from the host's meal. Some species (there are ten, ranging in size from several inches to three feet) may supplement these meals by nibbling on parasitic crustaceans attached to their hosts.

Vampire lamprey. *Above, three lampreys have about finished draining the blood and body fluids from a salmon. The salmon will, of course, die.*

Weakener. *The isopod attached to the sea bass, at right, weakens the bass to a point where it can no longer defend itself against other predators.*

Parasitism

Lampreys are the vampires of the sea. Above we see a salmon parasitized by several lampreys, which will literally drain the salmon of its lifeblood.

Lampreys are among the most primitive of fishes. They attach themselves with jawless, suckerlike mouths and scrape their victims' skin away with a rasplike tongue. They live in the sea and, like the salmon they often parasitize, spawn in fresh water. Lamprey-control programs, therefore, are usually carried out in the spawning streams. But lampreys aren't the only parasites on fish. The large isopod crustacean opposite is attached to a sea bass. It clings to its victim by biting with pincers, then pressing out blood and other body fluids.

The catch. *The octopus, seeing its favorite meal, a lobster, sends out one of its arms and attaches a suction cup to its tail.*

The meal. *Having hauled in the prey, the octopus crushes the shell of the lobster with its beak and enjoys its meal.*

The Feeding Octopus

Ghastly apparition of the deep—or shy, intelligent being? The octopus has an undeserved reputation reflected in one of its names—devilfish. Actually, we have found the octopus to be a retiring, unaggressive creature who uses his high intelligence to outwit enemies and find food. Here the octopus stealthily approaches a lobster, snaps it up, then stuffs it into his mouth with his tentacles. Lobster and other crustaceans are a typical diet of octopods. Some grow to 30 feet across and weigh up to 125 pounds. Others never exceed a few inches. When we were filming them in the Mediterranean we found they could quickly jet away from us as a defense.

Predators That Follow Predators That Follow Predators

Tuna, one of the fastest and hungriest of open-sea fishes, tour the world in unwearying pursuit of potential dinners. Many of the great tuna schools are accompanied by herds of dolphins—equally fast, equally strong, but far more intelligent—who are looking out for their own diet. In turn, many sharks follow, at a safe distance, the dolphins. Man

also hunts the tuna and takes hundreds of thousands of them a year in his huge nets. Unfortunately, intelligent as dolphins are known to be, for some reason they have not learned to make themselves scarce when a fishing fleet bears down on a school of tuna.

A pod of dolphins. *Seen here is a school, or "pod," of dolphins leaping out of the water and traveling in what seems to be a "V" formation.*

The terrible result is that about 200,000 dolphins accidentally perish each year when they are caught up in tuna nets.

Chapter XI. Primitive Fishing Methods

Fishing is not limited to the catching of fish. It includes the capture of any marine animal. The earliest fisherman used no tools to assist him. He merely waded through the shallow waters of streams, lakes, and oceans near his home, picking up snails, mussels, crabs, and anything else he found. Small pools left by the receding tide, or ponds drying up, were the best places. In the enclosures fish had little room to hide and could be captured with bare hands. In streams salmon often remain nearly stationary with respect to the shore, and a stealthy and skilled hunter could slip his hand along the fish's body from behind, grasping the gill openings and tossing it up onto the bank.

"The basic challenge that fishermen face is being unable to see well beneath the surface. They must rely on remote systems to indicate the presence of fish."

Fishing consists of two basic phases: location of prey and capture of it. Fishermen today use scientifically advanced but still inefficient methods of locating prey; however, they have improved their methods of capture. The basic challenge that fishermen face is being unable to see well beneath the surface. They must rely on remote systems to indicate the presence of fish or conditions optimum for fish. Fishermen, in the past as well as the present, rely on other animals, especially birds, to indicate the presence of schooling fishes. Certain birds are found where certain types of fish are active.

The spear, or fish stick, was invented hundreds of times in the prehistoric past by fishermen thousands of miles apart. It started with a straight stick with a pointed tip, then evolved to a barbed tip which keeps the stabbed fish from falling off the shaft. Spear points were fashioned of bones, shells, small sticks, or sharpened rocks. The sea god Poseidon carried a three-pronged trident, the fish spear of ancient Greeks. Diving fishermen in Polynesian South Sea islands polish tortoise shell to near-perfect transparency to use as face masks in undersea hunting.

The advent of basic tools increased the efficiency of fishermen. Sticks and rocks were used to break the grip shellfish like oysters have on rocks. Octopuses dwelling in shallow holes could be prodded from their hiding places with long sticks. Rafts and floating logs, eventually boats or canoes, enabled the fishermen to venture away from the shore. As he moved farther out to sea, shellfish were discovered in deeper waters and men and women began diving to collect them. In Japan diving women called *ama* still dive as their ancestors did more than 2000 years ago, without mechanical assistance.

Chemical means were (and are still) used to capture fish. Fish can be stupefied by introducing the juices of certain plants to small bodies of water where they will not diffuse rapidly. The chemicals act like an anesthetic, probably affecting a fish's motor nerves and muscles, so that it floats to the surface. Another way men fish with chemicals is by poisoning small fish and then releasing them into the water. The erratic swimming motions make them easy prey for larger fish, which then succumb.

Spearing. *At right, a fisherman practices his trade in the most primitive of ways. He spears the fish—a grouper, in this case.*

Primitive Ingenuity

Scoop baskets and cover pots were once used to catch fish in shallow water. Fashioned out of reeds, bamboo, or branches lashed together, they resembled giant bowls. Fishermen rested scoop baskets on the bottom. When a fish approached, the scoop was quickly raised. Cover pots were similar but had a small hole in the top. They were most effective when many men fished together in muddy shallows. The fishermen simultaneously dropped their cover pots to the bottom, large hole down, trapping any marine life resting there or passing underneath. The fish-

> "Fishermen's ingenuity is not limited to thinking of new ways to capture food."

erman could then reach through the small hole in the top of his basket-line trap to feel if anything had been caught. Some fishermen used woven lift nets which they spread on the bottom. When a fish passed over, a quick lift of the net often resulted in capture.

The fisherman's ingenuity was not limited to thinking of new and more efficient ways to capture food. Once his scoop or net was filled, he needed a place to put the catch. Some of

Native fishermen on Lake Rudolf. *Around the margins of one of East Africa's most beautiful lakes, work equals fun—as men pit their wits and wicker traps against the swiftness of their catch.*

the Pacific islanders solved this problem by digging holes in the earth near shore. When the hole was completed, the fishermen splashed water into it until it was partly filled. The pool could then be used to keep turtles and fish alive.

Chapter XII. Fishing the Sea Today

Individual fishermen using simple techniques to feed themselves and their families never threatened a marine species with extinction. Their methods are too inefficient. But when steam power was introduced to the fishing fleets of the world during the 1800s we began to take the first steps away from the primitive fishing techniques we had used for so long. Steam engines allowed fishermen to fish equally effectively in times of calm and of storm. And the power was harnessed to deck machinery, so fewer men were needed to haul nets and trawls.

During World War II whole fishing fleets were destroyed; nations and individuals had to start rebuilding from scratch. They incorporated all the technological advances made during the war years into their fleets —electronic detection and communication gear of all kinds. Sounding equipment was used to detect schools and thermometers guided the new fishermen on the trail of migrations. Radio telephones were used to keep contact with home offices and helicopter-borne spotters. Synthetic materials appearing in the late 1940s also had a profound effect on the industry. Fishing nets made of these are lighter and more durable than those made of natural fibers. However perfected modern fishing has become, it still remains essentially as barbarian as hunting was for the caveman. It consists of preying upon existing stock with no consideration for depletion of that stock.

Traditionally man has been happy-go-lucky in his fishing of the sea. He sets sail from his home port, goes to the spot where long experience tells him a catch will be found, throws over his nets, pulls up whatever he can, and returns to harbor. He has almost never had to worry about the next day's or next year's catch; the ocean was more than wide and deep enough to ensure that. Only rarely has an alteration in the life of the sea undermined his assumptions, as when a change in the migratory habits of Baltic herring in the late Middle Ages disrupted the politics of Northern Europe. Fishing on the scale, and with the techniques, of today is something else again. All over the world immemorially rich grounds are thinning out; governments of countries dependent upon fishing are quarrelling over what is left.

Today highly organized fleets of vessels also systematically hunt fish and whales. The factory-ship system, with a single large vessel coordinating fishing activity of smaller vessels, is highly effective. Heavy fishing by such fleets has dramatically reduced the populations of cod, yellowtail flounder, and haddock.

" Modern fishing remains as barbarian as hunting was for the caveman—preying upon existing stock with no consideration for depletion of that stock."

To avoid bankrupting the sea we must learn the influence each species of plant and animal has upon its neighbors, for the extinction of a single species could doom others dependent on it. For example, the overfishing of tuna results in the proliferation of the smaller bonito, which in turn eats more tuna eggs and larvae, accelerating the depletion of tuna.

Haul of sole. *Modern seagoing fishing vessels are capable of catching huge numbers of fish. All too often these numbers are more than the sea can afford.*

A Dying Industry

The last of the whales? Not quite, but as the whale population of the world has declined, so has the whaling industry. Only South Africa, Japan, and USSR have active whaling fleets today. Norway and the U.S.A. recently gave them up. A hundred years ago, hundreds of whaling ships poured forth from the ports of the seafaring nations to seek the oil and flesh of the whale. They harpooned them by hand and administered the coup de grace with long, sharp lances, which pierced the whale's heart as the animal fled men in small boats. Later the harpoon was mechanized and powered with explosives. The vessels became swift catcher craft. Airplanes spotted the whales and radioed messages to the catchers. And the whales succumbed. Today the populations of some species are so small that experts wonder if males and females can even find each other in the vastness of the world's oceans. The International Whaling Commission, meeting yearly, sets quotas and seasons when whales may be hunted. They delineate areas from which

Harbor facilities. *In the photo above, at a South African pier, two whales are readied for rendering into commercial products.*

whales may be taken and areas where they are protected. But the enormous size of the whaling grounds makes enforcement difficult. And the two major whaling nations do not belong to the commission.

Another fishery which is in the midst of change is the Peruvian anchovetta industry. Each year Peruvian fishermen net close to 20 million tons of these small herringlike fishes close off their native shores. The fish are processed into somewhat more than 2 million tons of fish protein concentrate, the fish meal so sorely needed by starving peoples of the world. But when periodic changes in the nutrient-laden Humboldt Current, which runs up the west coast of South America, disperse the anchovetta, the industry suffers severely. The related guano industry, dependent on anchovetta-eating guano birds of the Peruvian coast, also suffers badly.

Research vessel. *Above, scientists aboard a research vessel toss plankton nets overboard. The samples will tell them about local populations.*

Factory ship and trawler. *The trawlers in a fleet do the actual fishing, then bring the harvest back to the larger factory ship to be processed.*

Sophisticated Fishing Methods

Modern methods have increased the world's annual fish catch spectacularly in recent years. Where once trawlers and seiners brought their hauls back to port for processing, now many travel a few miles to a factory ship, shown here with a fishing boat. Research scientists are constantly seeking new ways to find, attract, and catch fish. Using paired plankton nets that look like and are called bongos, they take plankton samples. The whereabouts of plankton tell them where to find different species of fish.

Modern Fishing Demands

The exploding population of the world in recent years has placed unprecedented demands on the fishing industry. In a recent seven-year period — 1964 to 1970 — the catch of fish, shellfish, and crustaceans rose nearly 75 percent. As the table shows, the catch of salmon, trout, smelt, and related species tripled in that period. The catch of cod, haddock, hake, and pollack nearly doubled.

> "From 1962 to 1971 Peru maintained the lead as the world's main fishing nation."

These increases are due largely to innovations and improved fishing techniques initiated by the Japanese and the Russians, plus the fantastic growth of the Peruvian fishing industry. In the United States most trawlermen are still sailing in beam trawlers, considered far less efficient than the stern trawlers the Russians use. Stern trawlers require smaller crews, make more tows per day, and reap larger catches. Japanese innovations helped that nation hold rank as the world's leading fishing nation for many years. As late as 1961 Japan's take of more than 14 billion pounds kept it on top. By 1962, even though Japanese fishermen had increased their catch by more than a billion pounds, Peru took over first place. From 1962 to 1971 Peru maintained its lead, catching almost 28 billion pounds of fish in 1970 alone. Virtually the entire Peruvian fishery was for anchovetta, the herringlike fish that normally occurs in abundance in the cold, nutrient-filled Humboldt Current.

With the increase in the world's population, even the tremendously increased catches through discovery of new fisheries and use of more modern techniques are unable to keep up with the need. What will happen when, as the population continues to increase dramatically, world fisheries begin to decline from overfishing? Already the catches of some commercial species has declined in some regions. Haddock and yellowtail flounder catches in New England are down as Russian and Spanish fishing fleets work George's Bank. The anchovettas may be declining in Peruvian waters.

Serving a rich fishery. *This Icelandic harbor plays host to many fishing vessels. Two are seen in the foreground.*

Fishery Product	1964	1970
	Millions of pounds, live weight	
Herring, sardines, anchovies	40,760	48,505
Cod, hake, haddock, pollack	13,320	23,669
Molluscs	5,870	7,275
Redfish, bass, congers	6,510	8,347
Tuna, bonito, billfish	3,040	3,840
Salmon, trout, smelt	1,300	4,608
Crustaceans	2,560	3,571
Flounder, halibut, sole	2,160	2,826
Total	75,520	102,641

FRÓÐI RE. 44
SKÓGAFOSS RE236

Polluting Our Waters

In this picture the fact that the outfall is largely a dye only dramatizes the nature of the stuff our industrial civilization dumps into the ocean every day. Sewage outfalls near major cities can release as many as 300 million or more gallons of treated water per day *apiece!* "Treated" does not necessarily mean that the contribution is harmless. It may, and often does, profoundly modify the adjacent ecosystems: killing some forms, giving stimuli to others, setting in motion consequences no one paid any attention to. In *Oasis in Space* we saw how outfalls in California transformed a creative ecosystem into a sterile one: the sea urchin, thriving on sewage, pushing out its ecological "partner," the kelp, from vast areas of the ocean floor. The answer: more understanding of the biology of the sea in order to render all wastes as harmless as possible.

Food from the Sea

Many of the finest foods come from the sea: turbot, turtle soup, whiting, a hundred varieties of clam, Maine lobster—the list is long. We tend to forget that the bulk of the 60 million tons of fish taken out of the sea annually is less exotic but more essential. When traditional fishing patterns are disrupted, or when the potential productiveness of the sea through mariculture is compromised by pollution, not only are men thrown out of work and regions impoverished, but other men, women, and children will go hungry. All world food reports for the past 20 years have emphasized the growing gap between men who eat well and men who don't. It is as if the human race were dividing into two parts: those whose diets give them strong minds and bodies and those whose diets don't. One answer: global management in exploiting the ocean.

Chapter XIII. Mariculture

When the savants of the world gather at such meetings as the Stockholm conference on the environment, each of them is haunted by the same nightmare: by the year 2040 the population of the earth may have passed 16 billion, five times what it was in 1970. Where is the food to come from to feed these new billions?

> "It is wasteful to seek large predators. We could avoid the losses due to this inefficiency of conversion by seeking food nearer the base of the pyramid."

Up to now man has simply gone forth upon the sea and taken from it what he could—with the result that some species are showing serious effects of overfishing. We are also damaging the ocean by dumping into it a wide variety of pollutants—many of which disrupt the natural marine food chain of what we call the "living broth" of the sea. We have seen earlier in these books that all sea life participates in a great symbiotic relationship: extended "mutualism" in which each organism depends upon another in the web cf life. Even those preyed upon contribute to the balance of nature. Man is, of course, part of this grand relationship, but has more and more become a true "parasite." He may even be degenerating into a newly evolved global parasite that consumes its host to the point where the host cannot survive; this obviously results in the death of the parasite as well. But man has a brain: all this can—and will have to—change.

One of the basic ideas in farming the sea is to increase efficiency. Because there may be as much as a 90 percent loss between weight gained and food consumed at each step in the food pyramid, it is wasteful to seek large predators. We could avoid the losses due to this inefficiency of conversion by seeking food nearer the base of the pyramid. If man wants to harvest the sea, he will obtain a greater yield by exploiting animals which graze on plankton or on plant life.

The English ecologist Lord Ritchie-Calder has suggested the construction of an artificial whale, like a weird submarine with an open mouth, that could forge through the krill of the plankton layers and process tons of little crustaceans into edible protein. Another basic idea is to increase the productivity of given areas of the ocean either by using fertilizers or selected wastes or by creating artificial "upwelling" currents from depths of 2000 to 3000 feet. This upwelling would bring rich deep waters into the sunlit layers where the plankton thrives—and consequently the larger animals (in an experiment in the Virgin Islands it was discovered the phytoplankton grows 27 times as fast in water pumped from the deep as from the surface).

Agencies or individuals have also proposed to enclose areas of the ocean itself like the immemorial fishponds of many Asian countries, or to use helicopters and dolphins as sheep dogs, to transplant fish, to build protected hatcheries, to warm up fjords, to "ranch" with aqualung-equipped undersea farmers, to convert the Gulf of California into a huge fish farm, etc. But all these zealous proposals may be premature in our present state of marine ecological ignorance. Converting great expanses of coastlines to naive farming may be just as harmful as carelessly constructing marinas.

Improving on Nature

Fantastic dream. *A grandiose scheme for one day enclosing the Gulf of California, above, and exploiting it as a giant underwater farm has been proposed.*

One futuristic mariculture scheme is the placement of an electronic-screen "dam" across the access to the sea of the Gulf of California, converting this immense body of water into a gigantic fishpond. In Taiwan, where milkfish ponds are fertilized, the average annual yield is 520 tons per square mile. In other Asian countries where sewage is used instead of commercial fertilizer, the yield is far higher. But as Gifford P. Pinchot says, "The United Nations Food and Agriculture Organization has calculated that more than 140,000 square miles of land in southern and eastern Asia could be added to the area already devoted to milkfish husbandry. Even if this additional area were no more productive than the ponds of Taiwan, its yield would be more than today's total catch from all the world's oceans."

Platforms for Abalones

Abalone has been called the "steak of the sea." Although it is tough and rather tasteless, it is an expensive "luxury" food. Farming of abalone on the west coast of the United States is still in experimental stages, but marketable abalone will probably be raised by the mid- or late 1970s. Thus far scientists have been able to hatch eggs of this relative of the snail in the laboratory. But mortality of the larvae is the major stumbling block. Typically, the abalone lays about 2 million eggs. When these hatch they remain as free-swimming larvae for about a week.

Farming the Pompano

For centuries man has raised animals on land. Now he is about to begin raising marine animals. One of the best fish for eating is the pompano, a member of the jack family. Here young pompano are being raised in long tanks until they reach a size large enough for transplanting into impoundments. Under these strictly controlled conditions, farmers can keep track of water quality to encourage rapid, healthy growth. And they can feed the fish carefully measured amounts of the best foods. Properly done, pompano farming speeds the growth and increases the harvest.

Kelp cutter. *The machine above cuts the tops from the kelp, allowing the plants to stay alive. The tops furnish raw materials which make paint and ice cream smooth, and improve cosmetics.*

Octopus. *One of the chemical substances in the blood of the octopus is called* elecosin. *Researchers are hoping it may one day be useful in the control of high blood pressure.*

New Products from the Sea

Hopefully, researchers may one day find in the ocean answers to questions that will help to realize dreams of lengthening life and memory, of retarding aging, and of cures for a variety of illnesses. Organic anesthetics can be extracted from marine animals , eliminating the need to introduce synthetic chemicals into our bodies. Cephalotoxin is a poison found in the saliva of the octopus that hinders the coagulation of the blood. Elecosin is also found in the octopus's saliva; it may one day be used to control high blood pressure. The toxin used by stonefish may perhaps combat heart disease. The hagfish, a relative of the eellike lamprey, has four hearts, but only one is connected to the fish's nervous system. The beating of the others is stimulated by a chemical called eptatretin which someday may replace electronic pacemakers. One-celled planktonic organisms (dinoflagellates) give off a powerful poison that affects many fish and makes shellfish poisonous to humans. The poison reduces human blood pressure and depresses respiration. Because of its effect it may eventually be used to relieve hypertension. Some cone snails produce a powerful poison sometimes fatal to humans. The toxin from one species of these snails stimulates contraction in muscles, while that of a close relative causes muscles to relax. Extracts from the first may help restore diseased muscles.

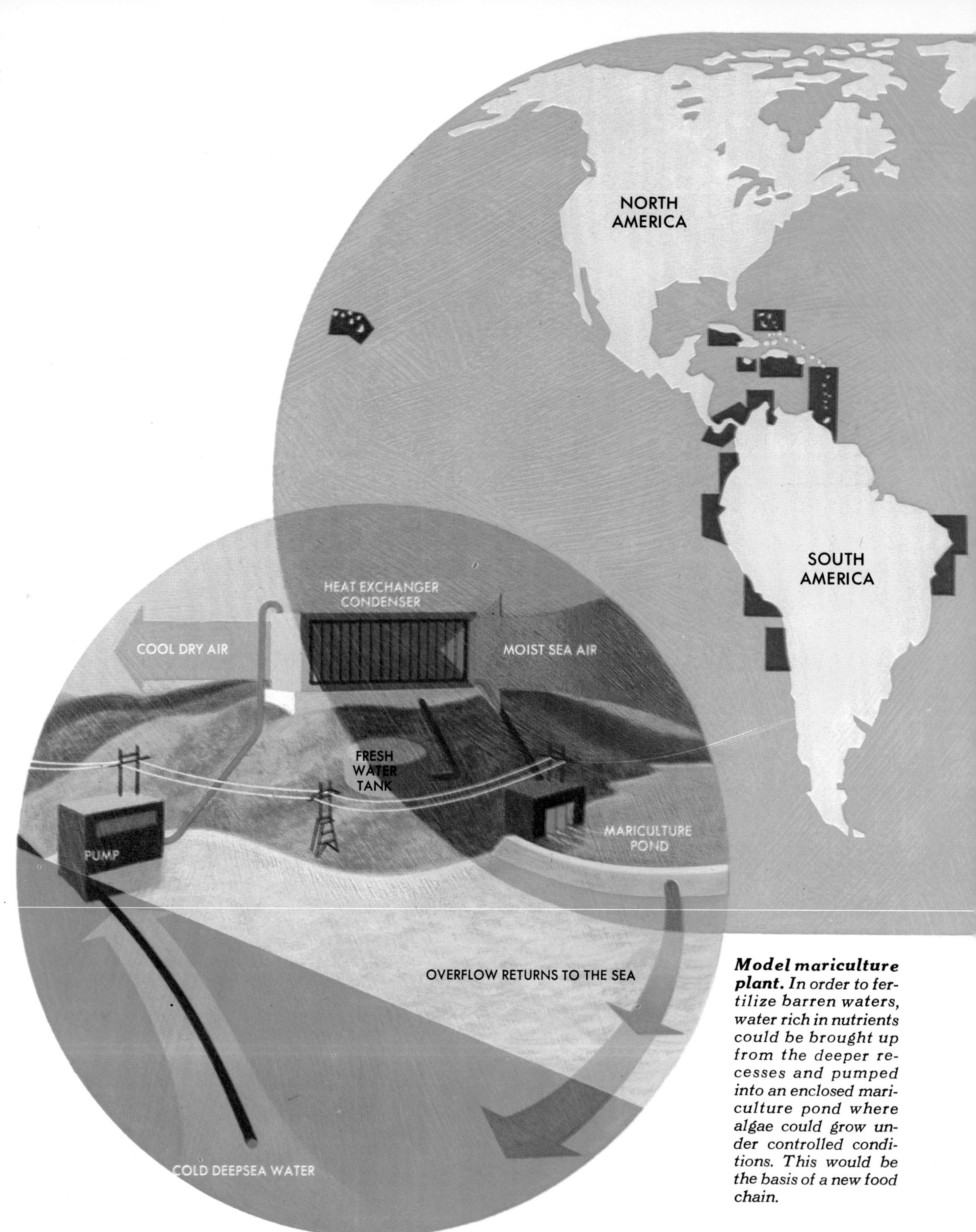

Model mariculture plant. *In order to fertilize barren waters, water rich in nutrients could be brought up from the deeper recesses and pumped into an enclosed mariculture pond where algae could grow under controlled conditions. This would be the basis of a new food chain.*

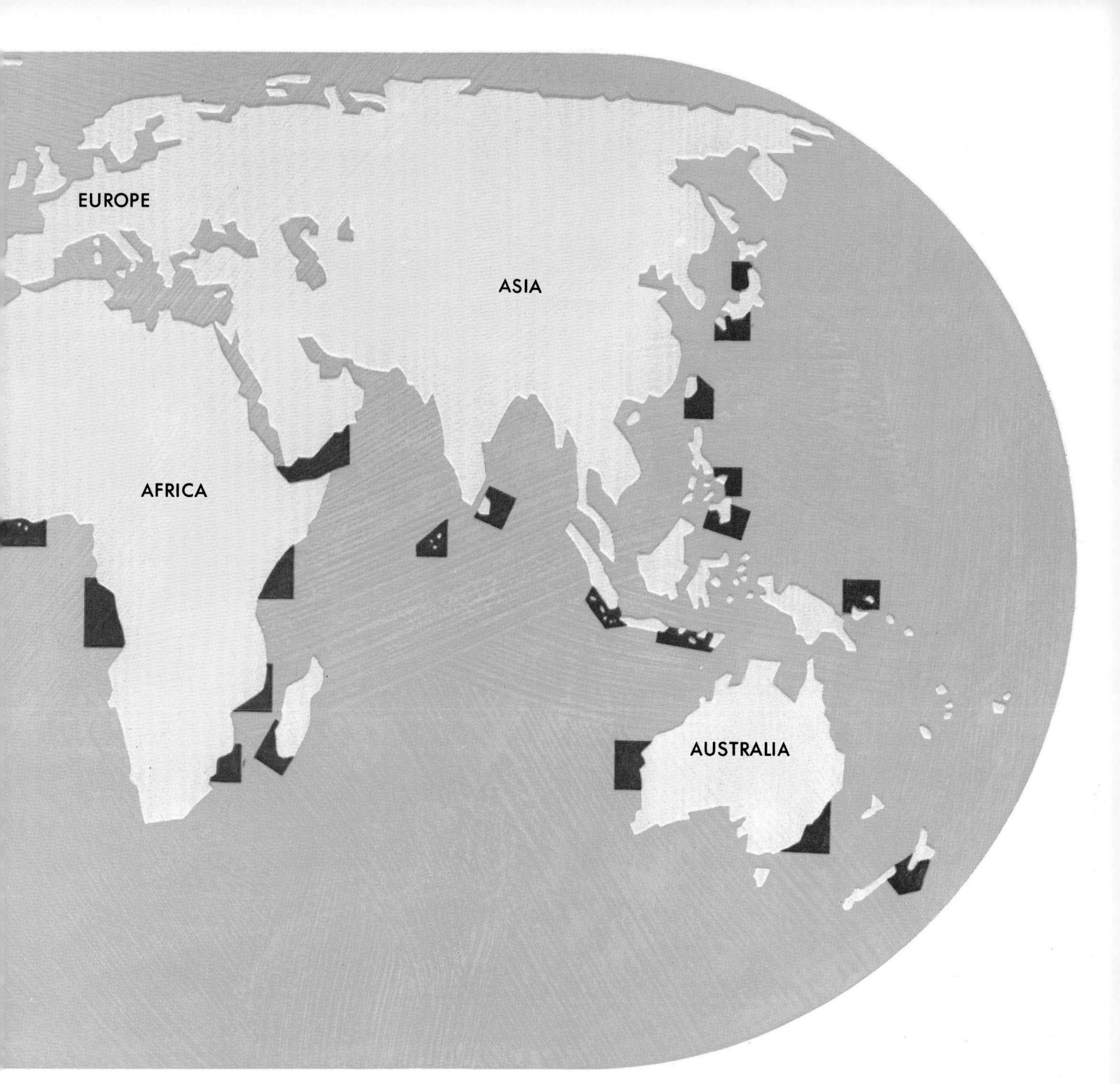

Artificial Upwelling

Farming the sea can mean many things from raising oysters to growing algae to feed fish. One method proposed for mariculture has been the drawing of nutrient-rich water from cold depths for the fertilization of sea farms. Areas that have been found suitable for this sort of farming are shown on the map. In the inset is a possible way of drawing the deep water up and pumping it into impoundments where fish can be raised.

Deep nutrient-laden water is pumped to a heat exchanger situated on a hilltop facing into the wind. Here it is warmed by the wind before it flows into an impoundment. A by-product of this system is fresh water condensed from the warm saturated sea winds as they pass through the heat exchanger and are cooled.

The Balance of Nature

Beneath the surges of the gales
Dumb waters ring with shattered hush
God's water fresh in heart and scales
And devil's beat ablaze in flesh.

When hunger reigns teeth turn acid
Wiles and fears run out of breath
Greedy shadows, eager, avid,
Fling in a dance of anti-death.

Drops of sea hurled far in the sky
As silver wakes break off to fly.

From nightmare brine squirt squids afire
For food orgies till nights expire.

Shining nuggets throb and crackle—
Swarming trillions living jewels
Lapse, engulfed by giant whales.
Danger arouses tentacles.
Sneaks dig their lair within their fear
Driving long saps that sea snakes clear.

Hunting flowers—toxic festoons
Electric venoms—fish cocoons.

Shreds of immunity
For a scarf of beauty.

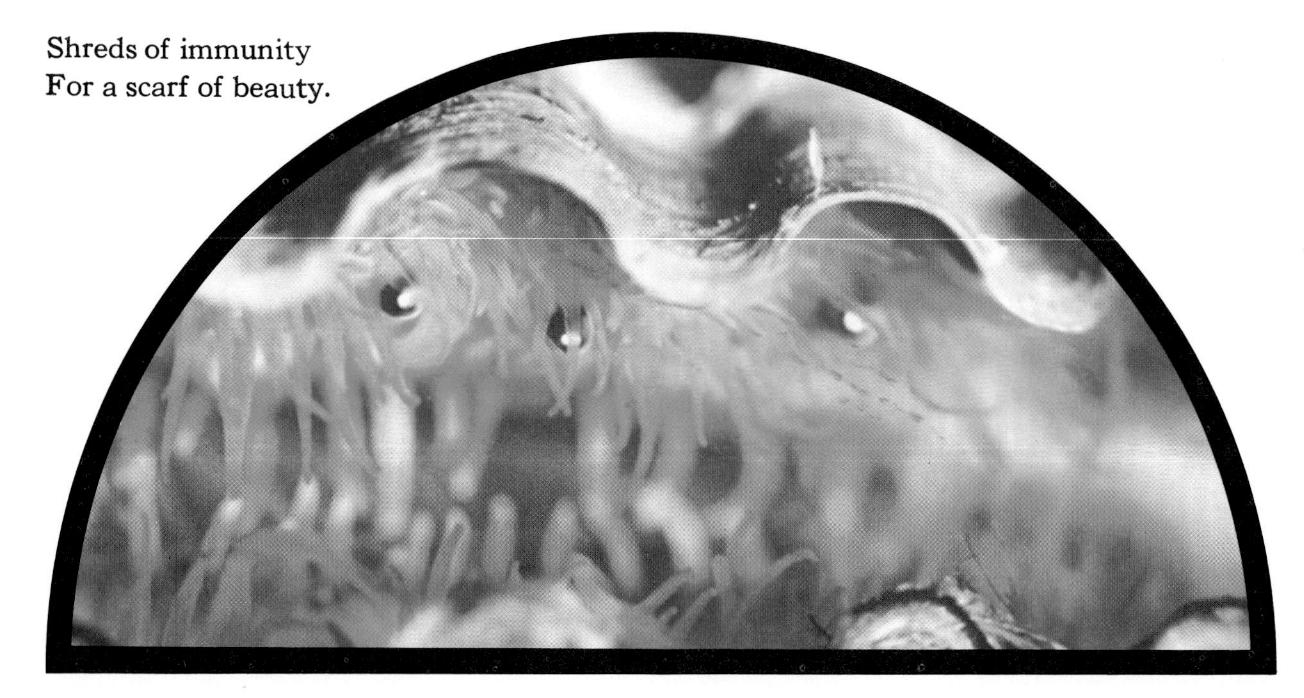

Index

A

Abalones, 20, 21, 136
Alcyonarians, 50
Algae, 28, 45, 50, 61, 84, 86
 green, 106
Amoebas, 22, 28
Anchovettas, 127, 130
Anchovies, 52, 130
Anemones, 22, 40, 104
Angelfish, 56
Anglerfish, 88
Arrow worms, 52

B

Baleen whales, 98–99, 134
Banded shrimp, 66
Barbels, 62–63, 88
Barnacles, 11, 31, 68, 74, 104, 111
Basking sharks, 16–17
Bass
 sea, 108, 114, 130
 striped, 37
Bat rays, 91
Bears, polar, 18, 20
Beebe, William, 8, 88
Billfish, 130
Bivalves, 68
Blacksmith fish, 107
Blennies, 27, 73
Bluefin tuna, 37, 40
Bluefish, 84, 94
Blue whales, 29, 99
Bongos, 128
Bonitos, 124, 130
Bristle worm, 34
Bryozoa, 70
Butterflyfish, 107

C

California gray whales, 35
Calypso, 10, 24, 65, 87, 100
Camouflage, 78–79
Cannibalism, 26
Catfish, 62, 65
 striped sea, 62
Cephalotoxin, 138
Chichlids, 32
Clams, 68, 80, 133
 tridacna, 44–45
Cleaners, 107–9
Clownfish, 104
Cod, 124, 130
Coelenterates, 76
Commensalism, 104, 113
Cone snails, 77, 138
Conger eels, 26, 130
Copepods, 52
Corals, 4, 40, 50, 70, 106
 alcyonarian, 50
 gorgonian, 50, 70
 organ-pipe, 50
 red-and-black, 50
Cover pots, 122
Crabeater seals, 42
Crabs, 20, 21, 66, 104, 108, 120
 hermit, 104
 stages of growth, 30
Crinoids, 74
Crustaceans, 27, 30, 50, 52, 66, 130

D

Defense, 78–79, 100, 117
Dinoflagellates, 17, 138
Dolphins, 18, 29, 32, 39, 40, 42, 43, 56, 58, 103, 118–19, 134
Dragonfish, 88, 89
Dugongs, 92
Dumas, Frederic, 112

E

Echolocation, 56, 58
Ecosystems, 132
Eels
 Anguilla, 17
 conger, 130
 garden, 73
 moray, 65, 108
Eggs, 32, 37
Elecosin, 138
Emperor penguins, 36
Energy, 18–21, 30, 31, 32, 36, 40, 42, 46, 52
 solar, 52
Eptatretin, 138
Eskimos, 29
Evolution, 4, 68, 92
Eyes, 60

F

Factory-ship system, 124, 128
Farming. *See* Mariculture
Feather-duster worms, 11, 68, 74
Feather stars, 74
Finback whales, 29, 99
Fish meal, 127
Fishing
 modern, 124–30
 primitive, 120–23
Fixed strainers, 74
Flatfish, 78
Flounder, 31, 130
 yellowtail, 124, 130
Food chain, 10, 11, 52–55

G

Galápagos Islands, 92
Garden eels, 73
Gill rakers, 94, 97, 98
Goatfish, 63, 107
Goosefish, 46
Gorgonians, 50, 70
Green algae, 106
Groupers, 24, 60, 87,
Gulls, 12
Guppies, 32

H

Haddock, 124, 130
Hagfish, 138
Hakes, 63, 130
Halibut, 130
Hammerhead sharks, 60
Hermit crabs, 104
Herring, 94, 124, 130
Humboldt Current, 127, 130
Hyacinths, water, 92
Hydroids, 77, 82

I

Iguanas, marine, 61, 92
International Whaling
 Commission, 126–27
Isopods, parasitic, 114

J

Jacks, 84
Jellyfish, 14, 52, 77

K

Kelp, 68, 132
Killer whales, 23, 103
King penguins, 36
Krill, 98, 99, 134

L

Lampreys, 114
Lateral line organs, 56, 60
Lizards, 61
Lobsters, 26, 46, 66, 117, 133

M

Manatees, 84, 92
Man-of-war, Portuguese, 77, 110
Manta rays, 94, 96, 104, 113
Mariculture, 133, 134–41
Marine iguanas, 61, 92
Marine worms, 34
Marlins, 18, 40
Microzooplankton, 52
Milk, whale and dolphin, 38–39
Milkfish, 135
Moray eels, 26, 65, 108
Mosquitoes, 22
Mudskippers, 61
Mussels, 68, 120
Mutualism, 104, 110, 134

N

Narwhal, 29
Nematocysts, 76, 77
Nomeus, 110
Nudibranchs, 68, 82
Nutrition, 22

O

Octopuses, 8, 60, 61, 117, 120, 138
Orcas, 23, 103
Organ-pipe coral, 50
Outfalls, 132
Overfishing, 134
Oxidation, 46
Oxygen, 22, 46
Oysters, 11, 27, 68, 120

P

Palolo, 34
Parasitism, 104, 114
Parrotfish, 4, 48
Penguins, 36, 103
emperor, 36
king, 36
Photosynthesis, 17, 45, 106
Phytoplankton, 11, 16, 52, 54, 55, 134
Pilotfish, 112–13
Pinchot, Gifford P., 135
Plankton, 11, 13, 16, 17, 29, 30, 46, 48, 50, 52, 68, 74, 84, 98, 128, 134
Polar bears, 18, 20
Pollack, 130
Pollution, 132–33, 134
Polyps, coral, 4, 9, 11, 50, 68, 106
Pompano, 137
Porgies, 102
Porpoises, 39
Portuguese man-of-war, 77, 110
Propulsion, 40–45

R

Rays, 90
bat, 91
manta, 94, 96, 104, 113
torpedo, 90
Red-and-black coral, 50
Redfish, 130
Remoras, 104, 113
Reproduction, 17, 32–39
Right whales, 99
Ritchie-Calder, Lord, 134

S

Sailfish, 40
Salmon, 17, 32, 37, 43, 65, 114, 120, 130
Salps, 14
Sardines, 52, 130
Sargassum fish, 79
Schooling, 13
Scoop baskets, 122
Sea anemones, 22, 40, 104
Sea bass, 108, 114, 130
Sea cucumbers, 49, 88
Sea fans, 11, 50
Sea gulls, 12
Seahorses, 32, 60
Sea lions, 32, 42, 46, 56, 58
Seals, 8, 23, 32, 46, 56, 103
crabeater, 42
Sea otters, 8, 18, 20, 21, 100
Sea pansies, 50
Sea pens, 50, 70
Sea plumes, 50
Sea robins, 67
Sea snakes, 60
Sea urchins, 20, 21, 40, 68, 82, 132
Sea whips, 50
Self-protection, 78–79, 100, 117
Sharks, 8, 9, 10, 18, 32, 55, 56, 64–65, 84, 94, 102, 104, 112–13
basking, 16–17
hammerhead, 60
tiger, 64
whale, 97
Shipworms, 80–81
Shrimp, 108
banded, 66
Skipjack tuna, 40
Smell, sense of, 64–65
Smelt, 130
Snails, 42, 68, 77, 120
cone, 77, 138
triton, 82
Sole, 124, 130
Spears, 120
Sperm whales, 10, 56, 88
Sponges, 22, 70, 88
Squids, 56, 60, 88, 104
Starfish, 19, 42, 46, 51, 68
Stomiatoids, 88
Stonefish, 138
Striped bass, 37
Striped sea catfish, 62
Sturgeon, 62–63
Surgeonfish, 86
Swordfish, 40
Symbiosis, 104, 106, 134

T

Teeth, 84, 86, 88, 97
Teredo navalis, 80–81
Tiger sharks, 64
Torpedo rays, 90
Tridacna, 44–45
Triton snails, 82
Trophic level, 52
Trout, 130
Trumpetfish, 79
Tunas, 10, 11, 18, 21, 40, 42, 55, 84, 94, 118–19, 124, 130
bluefin, 37, 40
skipjack, 40
Turbot, 133
Turtles, 61, 104, 123

U

Upwelling, artificial, 134, 140–41

V

Vision, 60–61

W

Walruses, 100
Water hyacinths, 92
Whales, 10, 24, 26, 29, 32, 38, 58, 84, 98, 103, 111, 124, 126–27
baleen, 98–99, 134
blue, 29, 99
California gray, 35
finback, 29, 99
killer, 23, 103
orca, 23, 103
right, 99
sperm, 10, 56, 88
white beluga, 29
Whale sharks, 97
Whaling, 126–27
White beluga whales, 29
Whiting, 133
Worms, 88
arrow, 52
feather-duster, 11, 68, 74
marine, 34
shipworms, 80–81
Wrasses, 73, 108

Y

Yellowtail flounders, 124, 130

Z

Zooplankton, 50, 52, 54, 60, 99

ILLUSTRATIONS AND CHARTS:

Sy and Dorothea Barlowe—53, 94; Howard Koslow—43, 54, 55, 89, 98, 99, 140, 141.

PHOTO CREDITS:

Harold E. Allport, Jr.—16; John Boland—57, 66, 67, 70, 83, 87; Robert K. Brigham, National Marine Fisheries Service—95, 128–129 (top); California Department of Fish and Game: Jim Phelan, Jack W. Schott—25; Jim Cooluris—90; Ben Cropp—48, 62–63, 64–65, 97, 102, 105, 121; David Doubilet—26, 72; Jack Drafahl, Brooks Institute—2–3, 12–13, 19, 49, 71, 78–79; Freelance Photographers Guild: A. Garcia—135, Bob and Ira Spring—131, J. W. Thompson—27, J. Zimmerman—33; George Green—15; Edmund Hobson—107; Hyperion Sewage Plant—17; Steve Leatherwood, Courtesy Naval Undersea Center, San Diego—35; Don Lusby, Jr.—29; Tom McHugh, Marineland of the Pacific—24; Marineland of Florida: Nat Fain—38–39; Bud Meese—73; National Marine Fisheries Service—125; Chuck Nicklin—23, 91, 109, 136, 138; Warren Rathjen, National Marine Fisheries Service—129 (bottom); Bruce H. Robison—41; Carl Roessler—82, 115; Dr. Ronald C. Rustad, Case Western Reserve University—28; DeBoyd Smith—30, 31; Larry Stoltz—14; Jack Swedberg—132; Paul Tzimoulis—85, 96; U.S. Navy Photographs—36; Wards Natural Science Establishment, Rochester, N.Y., and Monterey, Calif—76, 142; Wometco Miami Seaquarium—80–81, 110.